SIXTH EDITION

EXERCISES IN HELPING SKILLS

A Manual to Accompany *The Skilled Helper*

SIXTH EDITION

EXERCISES IN HELPING SKILLS
A Manual to Accompany *The Skilled Helper*

Gerard Egan
Professor Emeritus
Loyola University of Chicago

Brooks/Cole Publishing Company

I(T)P® An International Thomson Publishing Company

Pacific Grove • Albany • Belmont • Bonn • Boston • Cincinnati • Detroit • Johannesburg • London
Madrid • Melbourne • Mexico City • New York • Paris • Singapore • Tokyo • Toronto • Washington

Sponsoring Editor: *Eileen Murphy*
Editorial Associate: *Susan Carlson*
Marketing Representative: *Thomas L. Braden*
Marketing Team: *Romy Taormina, Jean Thompson, and Deanne Brown*
Text Formatting: *Erick and Mary Ann Reinstedt*
Production Coordinator: *Mary Vezilich*
Printing and Binding: *Courier Stoughton*
Cover Design: *Katherine Minerva*

For more information, contact:

BROOKS/COLE PUBLISHING
511 Forest Lodge Road
Pacific Grove, CA 93950
USA

International Thomson Editores
Seneca 53
Col. Polanco
11560 México, D. F., México

International Thomson Publishing Europe
Berkshire House 168-173
High Holborn
London WC1V 7AA
England

International Thomson Publishing Japan
Hirakawacho Kyowa Building, 3F 418
2-2-1 Hirakawacho
Chiyoda-ku, Tokyo 102
Japan

Thomas Nelson Australia
102 Dodds Street
South Melbourne, 3205
Victoria, Australia

International Thomson Publishing Asia
221 Henderson Road
#05-10 Henderson Building
Singapore 0315

Nelson Canada
1120 Birchmount Road
Scarborough, Ontario
Canada M1K 5G4

International Thomson Publishing GmbH
Königswinterer Strasse
53227 Bonn
Germany

Printed in the United States of America

5   4   3   2   1

ISBN 0-534-35057-7

CONTENTS

Section 5   BASIC EMPATHY 24

Section 6   THE ART OF PROBING AND SUMMARIZING 41

PART THREE: STAGE I OF THE HELPING MODEL AND ADVANCED COMMUNICATION SKILLS 58

Section 7   STEP I-A: HELPING CLIENTS TELL THEIR STORIES 58

Section 8   RELUCTANT AND RESISTANT CLIENTS 69

Section 9   STEP I-B: I. THE NATURE OF CHALLENGING– HELPING CLIENTS CHALLENGE THEMSELVES 72

Section 10   STEP I-B: II. SPECIFIC CHALLENGING SKILLS 81

INTRODUCTION TO THE EXERCISES

The exercises in this manual are meant to accompany the sixth edition of *The Skilled Helper* by Gerard Egan (Brooks/Cole Publishing Company, Pacific Grove, California, 1998). The six parts of the manual—I. Laying the Groundwork, II. Basic Communication Skills for Helping, III. Stage I of the Helping Model and Advanced Communication Skills, IV. Stage II: Helping Clients Determine What They Need and Want, V. Stage III: Helping Clients Work for What They Need and Want, and VI. The Action Arrow: Making it all Happen—correspond to the six parts of the text. The "sections" of the manual correspond to the chapters of the book. The exercises themselves are numbered consecutively throughout the text.

A TRAINING PROGRAM FOR HELPERS

As the title *The Skilled Helper* suggests, helpers-to-be need, not just a helping model, but a range of techniques and skills to make the model work. They include basic and advanced communication skills, establishing working relationships with clients, helping clients challenge themselves, problem clarification, goal setting, development of an action plan, the implementation of the plan, and ongoing evaluation. These are creative and humane ways of being with clients in their efforts to develop a program of constructive change. The only way to acquire these skills is by learning them experientially, practicing them, and using them until they become second nature. The following are standard steps in a skills-training program:

1. **Cognitive Understanding.** Develop a cognitive understanding of a particular helping method or the skill of delivering it. You can do this by reading the text and listening to lectures.

2. **Clarification.** Clarify what you have read or heard. This can be done through instructor-led questioning and discussion.

*The desired outcome of Steps 1 and 2 is **cognitive clarity**.*

3. **Modeling.** Watch experienced instructors model the skill or method in question. This can be done "live" or through films and videotapes.

4. **Written Exercises.** Do the exercises in this manual that are related to the skill or method you are learning. The purpose of this initial use of the method or skill is to demonstrate to yourself that you understand the helping method or skill enough to begin to practice it. The exercises in this manual are a way of practicing the skills and methods "in private" before practicing them with your fellow trainees. They provide a behavioral link between the introduction to a skill or method that takes place in the first four steps of this training format and actual practice in a group.

*The desired outcome of Steps 3 and 4 is **behavioral clarity**.*

5. **Practice.** Move into smaller groups to practice the skill or method in question with your fellow trainees.

6. **Feedback.** During these practice sessions evaluate your own performance and get feedback from a trainer and from your fellow trainees. This feedback serves to confirm what you are doing right and to correct what you are doing wrong. The use of video to provide feedback is also helpful.

*The desired outcome of Steps 5 and 6 is **initial competence**
in using the model and in the skills that make it work.*

7. **Evaluating the Learning Experience.** From time to time stop and reflect on the training process itself. Take the opportunity to express how you feel about the learning program and how you feel about your own progress. While Steps 1 through 6 deal with the task of learning the helping model and the methods and skills that make it work, Step 7 deals with group maintenance, that is, managing the needs of individual trainees. Doing this kind of group maintenance work helps establish a learning community.

8. **Supervised Practice with Actual Clients.** Finally, when it is deemed that you are ready, apply what you have learned to actual clients. Supervision is an extremely important part of the learning process. Indeed, effective helpers never stop learning about themselves, their clients, and the helping process itself.

*The desired outcome of Steps 7 and 8 is **initial mastery**.*

The program in which you are enrolled may cover only a few of these steps. Comprehensive programs for training professional helpers, however, should eventually include all steps. Of course, the best helpers continue to learn throughout their careers—from their own experiences, from the experiences of their colleagues, from reading, from seminars, from formal and informal research, and, perhaps most importantly, from their interactions with their clients. Full mastery, therefore, is a journey rather than a destination.

THE PURPOSE OF THESE EXERCISES

Against the background what helping is all about, the exercises in this manual serve a number of purposes:

1. **A Behavioral Grasp.** They can help you develop a **behavioral** rather than merely a intellectual grasp of the principles, skills, and methods that turn helping models into useful tools. They give you a "feel" for methods and skills before you use them in interactions with others.

2. **Self-Exploration.** These exercises can be used to help you explore your own strengths and weaknesses as a helper. That is, they provide a way of having you apply the helping model to **yourself** first before trying it out on others. As such, they can help you confirm strengths that enable you to be with clients effectively and manage weaknesses that would stand in the way of helping clients manage problem situations.

3. **Personal Problem Solving.** You can use these exercises to become a better problem managers and opportunity developer in your own life.

4. **Client Participation.** You can help **clients** use these exercises selectively to explore and manage their own problems in living more effectively. These exercises provide one way of promoting client participation in the helping process.

5. **Training Clients in Problem Management.** You can help clients use these exercises to learn the **skills of problem management** themselves. Training in problem-management skills encourages self-responsibility in clients and helps make them less dependent on others in managing their lives.

DIFFERENT WAYS OF USING THESE EXERCISES

Academic Versus Experiential Programs. The way you use these exercises depends on the kind of training program you are in. The approach may be academic or experiential.

- **Academic Programs**. If the program is more or less academic in nature, then you can use these exercises to provide yourself with a *behavioral* feel for the model, methods, and skills involved in skilled helping. However, even if your program has an academic cast to it, you might want to get a learning partner and share the learnings you glean from these exercises. One way of learning the stages and steps of this helping model is to apply them to your own problems and concerns first. Applying what you are learning to the problems and unused opportunities in your own life is often a distinct benefit.
- **Experiential Training Programs**. If your program is experiential in nature, then these exercises can help you prepare for involvement with your co-learners. In experiential programs the instructor usually coordinates and integrates reading, lectures, the use of these exercises, and practice in trying out theh methods and skills of the helping model with co-learners.

Role Playing Versus Dealing With Personal Issues. There are various approaches in an *experiential* program. You may role play certain kinds of clients or you may work on your own real issues or you may do a combination of both.

- **Role Playing.** In experiential programs you are going to be asked to act both as helper and as client in practice sessions. In the written exercises in this manual, you are asked at one time or another to play each of these roles. Role playing is not easy. It forces you to get "inside" the clients you are playing and understand their concerns as they experience them. This can help you become more empathic.
- **Dealing With Real Issues**. Role-playing, while not easy, is still less personally demanding than discussing your own real concerns in practice sessions. Under certain conditions, the training process can be used to look at some of the real problems or concerns in your own life, *especially issues that relate to your effectiveness as a helper*. For instance, if you tend to be an impatient person—one who places unreasonable demands on others—you will have to examine and change this behavior if you want to become an effective helper. Or, if you are very nonassertive, this may assist you in helping clients challenge themselves.

Another reason for using real problems or concerns when you take the role of the client is that it gives you some experience of **being** a client. Then, when you face real clients, you can appreciate some of the misgivings they might have in talking to a relative stranger about the intimate details of their lives. Other things being equal, I would personally prefer going to a helper who has had some kind of personal experience as a client.

A Safe and Productive Training Group. If what you find out about yourself through these exercises is not to be shared with others, then dealing with real issues is not a concern. However, if you are going to talk about personal issues in the training sessions, you should do so only under certain conditions:

- **A Supportive Climate.** Dealing with personal concerns in the training sessions will be both safe and productive if you have a competent trainer who provides adequate supervision, if the training group becomes a learning community that provides both support and reasonable challenge for its members, and if you are willing to discuss personal concerns. Self-disclosure will be counterproductive if you let others extort it from you or if you attempt to extort it from others. You self-disclosure should always remain appropriate to the goals of the training group. Extortion, "secret-dropping," and dramatic self-disclosure are counterproductive.

- **Adequate Preparation for Self-Disclosure.** If you are to talk about yourself during the practice sessions, you should take some care in choosing what you are going to reveal about yourself. Making some preparation for what you are going to say can prevent you from revealing things about yourself that you would rather not discuss in a training group. Exercises to help you do just this are found later in this manual.

SOME CAUTIONS IN USING THIS MANUAL

First, it is important to note that the exercises suggested in this manual are means, not ends in themselves. They are useful to the degree that they help you develop a working knowledge of the helping model and acquire the kinds of skills that will make you an effective helper. *They are helpful to the degree that they help you begin to become proficient in helping clients achieve the overall goals of the helping process.* Other exercises can be added and the ones outlined here can be modified in order to achieve this goal more effectively.

Second, these exercises have been written as an adjunct to the text. They usually presuppose information in the text that is not repeated in the exercises. Also the examples used in the text will helped you do these exercises in a much more informed way.

Third, the length of training programs differs from setting to setting. In shorter programs there will not be time to do all these exercises. Nor is it necessary to do all of them. In shorter training programs, it is helpful to do at least one exercise from each of the sections on communication skills, one from each of the nine steps of the helping process, and one from the sections on action. This will help you develop a behavioral feeling for the **entire** helping model and not just the communication skills part of it. If the entire focus of the training program is on the communication skills that serve the helping process, you should understand that your training is incomplete. Further training in the helping model itself is needed.

Fourth, these exercises achieve their full effect only if you share them with a learning partner or the members of your training group and receive **feedback** on how well you are learning the model and the skills that make it work. Your instructor will set up the structure needed to do this. Since time limitations are always an issue, learning how to give brief, concise, behavioral feedback in a humane, caring way is most important.

> **Therefore, these exercises should be one part of a well-planned training program. It should never be a question of starting with the first exercise and then moving through all of them one after another. Rather the instructor-trainer should weave them into the fabric of an well-designed training program. This means selecting those exercises that best fit the program. Furthermore, time constraints very often demand a selective approach.**

The main complaints from students with regard to exercises are that they do not receive enough guidance in their use and that they are asked to do more exercises than is necessary and therefore feel overwhelmed.

PART ONE

LAYING THE GROUNDWORK

The exercises in this section relate to the first three chapters of the text. They lay the groundwork for the problem-management and opportunity-development helping process that is developed in *The Skilled Helper*. This includes outlining the nature and goals of helping (Section 1), providing an overview of the helping process (Section 2), and crafting the preferred culture of helping in terms of assumptions, beliefs, values, norms that should drive helping behavior (Section 3).

Section 1
INTRODUCTION

You may or may not intend to become a professional helper. Whether you do or not, learning the model, methods, and skills of *The Skilled Helper* can help you become more effective in your interactions with both yourself and others in all the social settings of life including family, friendship groups, and work settings.

The sections dealing with what helping is all about and the goals of helping are the most important part of Chapter One. It is important that you understand what helping is all about in order to get the most out of doing these exercises.

EXERCISE 1: UNDERSTANDING WHAT HELPING IS ALL ABOUT

1. Read the sections dealing with what helping is all about, the goals of helping, and helping as a learning process.
2. Picture yourself as helper talking to a prospective client.
3. In your own words describe to that person what you see helping to be.

_____

_____

_____

_____

_____

_____

_____

_____

_____

_____

_____

4. Share what you have written with a learning partner. Note the similarities and differences between the two statements. To what degree do they include the main ideas outlined in the text. In what way do you think you should modify your statement?

PROBLEMS AND OPPORTUNITIES
Preparing to Play the Role of Client

Here, in statement form, are some of the kinds of problems, issues, and concerns that trainees have dealt with during training programs. Remember that the flip side of every problem is an opportunity.

- I'm shy. My shyness takes the form of being afraid to meet strangers and being afraid to reveal myself to others.
- I'm a fairly compliant person. Others can push me around and get away with it.
- I get angry fairly easily and let my anger spill out on others in irresponsible ways. I think my anger is often linked to not getting my own way.
- I'm a lazy person. I find it especially difficult to expend the kind of energy necessary to listen to and get involved with others.
- I'm somewhat fearful of persons of the opposite sex. This is especially true if I think they are putting some kind of demand on me for closeness. I get nervous and try to get away.
- I'm a rather insensitive person, or so I have been told. I'm a kind of bull-in-the-china-shop type. Not much tact.
- I'm overly controlled. I don't let my emotions show very much. Sometimes I don't even want to know what I'm feeling myself.
- I like to control others, but I like to do so in subtle ways. I want to stay in charge of interpersonal relationships at all times.
- I have a strong need to be liked by others. I seldom do anything that might offend others or that others would not approve of. I have a very strong need to be accepted.
- I have few positive feelings about myself. I put myself down in a variety of ways. I get depressed a lot.

- I never stop to examine my values. I think I hold some conflicting values. I'm not even sure why I'm interested in becoming a helper.
- I feel almost compelled to help others. It's part of my religious background. It's as if I didn't even have a choice.
- I'm sensitive and easily hurt. I think I send out messages to others that say "be careful of me."
- I'm overly dependent on others. My self-image depends too much on what others think of me.
- A number of people see me as a "difficult" person. I'm highly individualistic. I'm ready to fight if anyone imposes on my freedom.
- I'm anxious a lot of the time. I'm not even sure why. My palms sweat a lot in interpersonal situations.
- I see myself as a rather colorless, uninteresting person. I'm bored with myself at times and I assume that others are bored with me.
- I'm somewhat irresponsible. I take too many risks, especially risks that involve others. I'm very impulsive. That's probably a nice way of saying that I lack self-control.
- I'm very stubborn. I have fairly strong opinions. I argue a lot and try to get others to see things my way. I argue about very little things.
- I don't examine myself or my behavior very much. I'm usually content with the way things are. I don't expect too much of myself or of others.
- I can be sneaky in my relationships with others. I seduce people in different ways—not necessarily sexual—by my "charm." I get them to do what I want.
- I like the good life. I'm pretty materialistic and I like my own comfort. I don't often go out of my way to meet the needs of others.
- I'm somewhat lonely. I don't think others like me, if they think about me at all. I spend time feeling sorry for myself.
- I'm awkward in social situations. I don't do the right thing at the right time. I don't know what others are feeling when I'm with them and I guess I seem callous.
- Others see me as "out of it" a great deal of the time. I guess I am fairly naive. Others seem to have deeper or more interesting experiences than I do. I think I've grown up too sheltered.
- I'm stingy with both money and time. I don't want to share what I have with others. I'm pretty selfish.
- I'm somewhat of a coward. I sometimes find it hard to stand up for my convictions even when I meet light opposition. It's easy to get me to retreat.
- I hate conflict. I'm more or less a peace-at-any-price person. I run when things get heated up.
- I don't like it when others tell me I'm doing something wrong. I usually feel attacked and I attack back.

While on the surface this looks like a list of problems, it also contains strengths that are either underused or misused. For instance, the person who talks about being "overly controlled" does have self-discipline. While the self-discipline may be misused or abused, it can still be a strength.

Therefore, one way of preparing yourself to talk about issues that can influence the quality of your helping is to review both strengths and soft spots in your life. All of us have strengths and soft spots, areas in which we could use some improvement. I might well be a decisive person and that is a strength for me as a helper but, in being decisive, I might push others too hard and that is a soft spot needing improvement. In this case, a strength pushed too hard becomes a soft spot.

EXERCISE 2: STRENGTHS AND SOFT SPOTS IN MY LIFE

Sometimes a very simple structure can help you and your clients identify the major dimensions of a problem situation. This exercise asks you to identify some of the things that are not going as well as you would like them to go in your life and some of the things you believe you are handling well. It is important right from the beginning to help clients become aware of their resources and successes as well as their problems and failures. Problems can be handled more easily if they are seen in the wider context of resources and successes.

1. In this exercise, merely jot down on separate paper—in whatever way they come to you—attitudes, skills, and behaviors that you have that help you lead a fuller life (strengths). Also list things that need some improvement (soft spots). In order to stress the positive, see if you can write down at least two things that are going right for each thing that needs improvement. Read the list in the example that follows and then do your own. Don't worry whether the problems or concerns you list are really important. Jot down whatever comes to your mind. Your own list may include items similar to those in the example, but it may be quite different because it will reflect you and not someone else.

Example

Strengths	Soft Spots
• I have lots of friends.	• I am extremely demanding of myself.
• Others can count of me.	• I get dependent on others too easily.
• I've got a good part-time job; they like the quality of my work.	• I am slow to take risks even when they are reasonable.
• I'm fairly smart.	• I can be arrogant with people who are not as smart as I am.
• I am very healthy.	
• My belief in God gives me a center.	
• I'm hard working.	
• I have a mind of my own; I don't give in just to please others	

2. Review your list once you are finished. Star one or two items in each column are the most important for you.
3. Share you list with a learning partner and discuss how you would like to make better use of the starred strengths and what you would like to do about the starred soft spots.
4. Keep this list. It can help you choose issues to discuss when you play the role of ciient.

Section 2
OVERVIEW OF THE HELPING MODEL

Here is a brief outline of the skilled-helper model. There is, of course, a fuller outline in Chapter Two of *The Skilled Helper*.

STAGE I: THE CURRENT STATE OF AFFAIRS—
Clarification of the Key Issues Calling for Change

Clients can neither manage problem situations nor develop unused opportunities unless they identify and understand them. Exploration and clarification of problems and opportunities take place in Stage I. This stage deals with the current state of affairs, that is, the problem situations or unused opportunities that prompt clients to come for help. This stage includes the following steps:

1. **Help clients tell their stories.** First of all, clients need to talk abut their problems and concerns, that is, tell their stories. Some do so easily, others with a great deal of difficulty. You need to develop a set of attitudes and communication skills that will enable you to help clients reveal problems in living and unused potential. This means helping clients find out what's going wrong and what's going right in their lives. Successful assessment helps clients identify both problems and resources. They also help clients spell out problem situations in terms of specific experiences, behaviors, and feelings.

2. **Help clients challenge their blind spots and develop new perspectives.** This means helping clients see themselves, their concerns, and the contexts of their concerns more objectively. This enables clients to see more clearly not only their problems and unused opportunities, but also ways in which they want their lives to be different. Your ability to help clients challenge themselves throughout the helping process can add a great deal of value.

3. **Help clients focus on issues that will make a difference**. This means helping clients identify their most important concerns, especially if they have a number of problems. Effective counselors help clients work on "high-leverage" issues, that is, issues that will make a difference in clients' lives.

STAGE II: THE PREFERRED SCENARIO—
Helping Clients Determine What They Need and Want

Once clients understand either problem situations or unused opportunities more clearly, they often need help in determining what they want instead of what they have. They need to develop a preferred scenario, that is, a picture of a better future, choose specific goals to work on, and commit themselves to a program of constructive change. For instance, at this stage a troubled married couple is helped outline what a better marital relationship might look like. Throughout the counseling process rusty client imaginations need stimulating.

1. **"What do you want?" Help clients develop a range of possibilities for a better future.** If a client's current state of affairs is problematic and unacceptable, then he or she needs to be helped to picture a new state of affairs, that is, alternate, more acceptable possibilities. For instance, for a couple whose marriage is coming apart and who fight constantly, one of the elements of the new scenario might be fewer and fairer fights. Other possible elements of this better marriage might be greater mutual respect, more openness, more effectively managed conflicts, a more equitable distribution of household tasks, and so forth. Separation or even divorce might be considered if differences are irreconcilable and if the couple's values system permits such a solution.

2. **"What do you really want?" Help clients translate preferred-scenario possibilities into goals.** Once a variety of preferred-scenario possibilities—which constitute possible goals or desired outcomes of the helping process—have been generated, it is time to help clients choose the possibilities that make most sense for them and turn them into an agenda, that is, a goal or a set of goals that form the first part of a constructive change program. A new scenario is not a wild-eyed, idealistic state of affairs, but rather a picture of what the problem situation would be like if improvements were made. The agenda put together by the client needs to be viable, that is, capable of being translated into action. It is viable to the degree that it is stated in terms of clear and specific outcomes and is substantive or adequate, realistic, in keeping with the client's values, and capable of being accomplished within a reasonable time frame.

3. **"What are you willing to pay for what you want?" Help clients commit themselves to the goals they choose.** Problem-managing goals are useless if they are not actively pursued by clients. They need to determine what they are willing to pay for a better future and then search for incentives to help them move forward. The search for incentives is especially important when the choices are hard. How are truants with poor home situations to commit themselves to returning to school? What are the incentives—the payoff, if you will—for such a choice? Most clients, like most of the rest of us, struggle with commitment.

STAGE III: STRATEGIES FOR ACTION—
Helping Clients Discover How to Get What They Need and Want

Discussing and evaluating preferred-scenario possibilities and choosing goals—the work of Stage II—determine the **"what"** of the helping process, that is **what** must be accomplished by clients in order to manage their lives more effectively. Stage III deals with **how** goals are to be accomplished. Some clients know what they want to accomplish, but need help in determining how to do it.

1. **Help clients brainstorm a range of strategies for accomplishing their goals.** In this step clients are helped to discover a number of different way of achieving their goals. The principle is simple: Action strategies tend to be more effective when chosen from among a number of possibilities. Some clients, when they decide what they want, leap into action, implementing the first strategy that comes to mind. While such a bias toward action may be laudable, the strategy may be ineffective, inefficient, uncreative, imprudent, or a combination of all of these.

2. **Help clients choose action strategies that best fit their needs and resources.** If you do a good job in the first step of Stage III, that is, if you help clients identify a number of different ways of achieving their goals, then clients will face the task of choosing the best set. In this step your job is to help them choose the strategy or "package" of strategies that best fits their preferences and resources. This tailoring of action strategies is important. One client might want to improve her interpersonal skills by taking a course at a college while another might prefer to work individually with a counselor while a third might want to work on his own.

3. **Help clients draw up a plan.** Once clients are helped to choose strategies that best fit their styles, resources, and environments, they need to assemble these strategies into a **plan**, a step-by-step process for accomplishing a goal. If a client has a number of goals, then the plan indicates the order in which they are to be pursued. Clients are more likely to act if they know what they need to do and in what order they do it. Plans help clients develop discipline and also keep them from being overwhelmed by the work they need to do.

CLIENT ACTION: THE HEART OF THE HELPING PROCESS
Helping Clients Act Both Within and Outside Counseling Sessions

Helping is ultimately about problem-managing and opportunity-developing **action** that leads to outcomes that have a positive impact on clients' lives. Discussions, analysis, goal setting, strategy formulation, and planning all make sense only to the degree that they help clients to act prudently and accomplish problem-managing and opportunity-developing goals. There is nothing magic about change; it is hard work. If clients do not act on their own behalf, nothing happens.

Two kinds of client action are important here: First, actions within the counseling sessions themselves. The nine steps described above are not things that helpers do to clients, rather they are things that clients are helped to do. Clients must take ownership of the helping process. Second, clients must act "out there" in their real day-to-day worlds. Problem-managing and opportunity-developing action is ultimately the name of the game. The stages, steps, and action orientation of the helping process make sense to the degree that lead to constructive change.

Since all the stages and steps of the helping process can be drivers of client action, action themes will be woven into each set of exercises. This will reinforce the principle that discussion and action go hand in hand.

CLIENT-FOCUSED HELPING

Helping usually does not take place in the neat, step-by-step fashion suggested by the stages and steps of the helping model. With some clients you will use only parts of the model just described. But you need the working knowledge and skill to be able to use whatever part is needed by the client. Effective helpers start wherever there is a client need. For instance, if a client needs support and challenge to commit himself or herself to realistic goals that have already been chosen, then the counselor tries to be helpful at this point. The nine steps of the helping model are active ways of **being with** clients in their attempts to manage problems in living and develop unused potential.

The needs of your clients and not the logic of a helping model should determine your interactions with them. One of the overriding needs of clients, of course, is to turn discussion and analysis of problem situations into problem-managing action. If you can help clients do this, you are worth your weight in gold.

Section 3
THE HELPING RELATIONSHIP: VALUES IN ACTION

It is important for you to take initiative in determining the kinds of values you want to pervade the helping process and relationship. Too often such values are afterthoughts. The position taken in Chapter Three of the text is that values should provide guidance for everything you do in your interactions with clients.

As indicated in the text, the values that are to permeate your helping relationships and practice must be owned by you. Learning about values from others is very important, but mindless adoption of the values promoted by the helping profession without reflection inhibits your owning and practicing them. You need to wrestle with your values a bit to make them your own. Second, your values must actually make a difference in your interactions with clients. That is, they must be values-in-use and not merely espoused values. Your clients will get a feeling for your values, not from what you say, but from what you do.
Note: Do the exercises in this section before reading Chapter Three in the text.

EXERCISE 3: WHAT IF I WERE A CLIENT?

1. Picture yourself as a client. Think of some of the problems you have had to grapple with or are struggling with now. Then picture yourself dealing with these with a counselor. Then ask yourself these questions. Jot down words, phrases, or simple sentences in response.

a. What would I want to get out of seeing a helper?

_____

_____

_____

_____

_____

b. What would I want the helper to be like?

_____

_____

_____

_____

_____

c. How would I want to be treated?

_____

_____

_____

_____

_____

2.   Share your statements with a learning partner. Discuss both the similarities and the differences.
3.   Finally, read Chapter Three in *The Skilled Helper*. That chapter deals with values that should permeate the helping process together with a preliminary consideration of the relationship between client and helper. Compare the values implicit in your answers to the above questions with the values outlined in Chapter Three. Share with a learning partner or the training group

EXERCISE 4: YOUR PRELIMINARY STATEMENT OF VALUES

1.   Write a preliminary STATEMENT OF VALUES based on your learnings from Exercise 1 that you as a helper could give to potential clients.

_____

_____

_____

_____

_____

_____

_____

_____

2.   Share your statement with a learning partner. Note both the similarities and the differences. Tell each other how you would feel, in light of the other's STATEMENT OF VALUES, what your hopes and fears would be were you to be your learning partner's client.

3.   Finally, read Chapter Three in *The Skilled Helper*. That chapter deals with values that should permeate the helping process together with a preliminary consideration of the relationship between client and helper. In the light of what you learn, revise your value statement and share it with a learning partner.

PART TWO

BASIC COMMUNICATION SKILLS FOR HELPING

There are two sets of communication skills that are essential for helpers. The first set includes attending, listening, basic empathy, probing, and summarizing is the focus of the exercises in Part Two. The second set, communication skills associated with helping clients challenge themselves, is dealt with in the exercises associated with Step I-B of the helping model—helping clients challenge their blind spots and develop new, action-oriented perspectives.

In Part II, Section 4 deals with attending and listening, Section 5 focuses on the communication of accurate empathic understanding, and Section 6 outlines the art of probing and summarizing.

Section 4
ATTENDING, LISTENING, AND UNDERSTANDING

Much of helping takes place through a dialogue between client and helper. If this dialogue is to serve the overall problem-management and opportunity-development goals of the helping process, the quality of the dialogue is critical. In this section the exercises focus on (I) your "attending" behavior and (II) active listening.

I. EXERCISES IN ATTENDING
Being Visibly Tuned In To the Client

Your posture, gestures, facial expressions, and voice all send nonverbal messages to your clients. The purpose of the exercises in this section is to make you aware of the different kinds of nonverbal messages you send to clients through such things as body posture, facial expressions, and voice quality and how to use nonverbal behavior to communicate with them. It is important that what you say verbally is reinforced rather than muddled or contradicted by your nonverbal messages. Before doing these exercises, read Chapter Five in *The Skilled Helper*, Communication Skills I: Attending and Listening. There are two points. First, use your posture, gestures, facial expressions, and voice to *send messages* you want to

14

clients, such as, "I want to work with you to help you manage your life better." Second, attend carefully so that you can *listen* carefully to clients.

EXERCISE 5: YOUR ATTENDING BEHAVIORS IN EVERYDAY LIFE

This is an exercise you do outside the training group in your everyday life. Observe your attending behaviors for a week—at home, with friends, at school, at work. You are not being asked to become preoccupied with the micro-behaviors of attending. Observe the quality of your presence to others when you engage in conversations with them. Of course, even being asked to "watch yourself" will induce changes in your behavior; you will probably "tune in" more effectively than you ordinarily do. The purpose of this exercise is to sensitize you to attending behaviors in general and to get some idea of what your day-to-day attending style looks like. First, read about attending skills in the text.

1.   Read the parts of Chapter Five, Attending and Listening, that deal with attending behavior.
2.   Watch yourself for a week attending (or not attending) to others in the various social settings.
3.   What are you like when you are with an listening to others, especially in serious situation? What do you do well? What needs improvement in your attending style?

Here is what one trainee wrote:

> I found myself attending better to people I like. When I was listening to someone one neutral, I found that my eyes and my mind would wander. It is easier for me to tune in to others when I'm rested and alert. When I'm physically uncomfortable or tired, I don't put in much effort to tune in. I was unpleasantly surprised to find out how easily distracted I am. However, simply by paying attention to "tuning in" skills, I was "with" others more fully, even with neutral people.

4.   Write a summary of what you have learned about yourself.

_____

_____

_____

_____

_____

5.   Share your summary with a learning partner and discuss.

EXERCISE 6: OBSERVING AND GIVING FEEDBACK ON QUALITY OF PRESENCE

In the training sessions, make sure that your nonverbal behavior is helping you work effectively with others and send the messages you want to send. Throughout the training program, observe the nonverbal behavior of your fellow trainees and give them feedback on how it affects you when you play the role of client or observer. Throughout the training program, ask for feedback on your own attending style.

Exercise 5, then, is an exercise that pertains to the entire length of the training program. You are asked to give ongoing feedback both to yourself and to the other members of the training group on the quality of your presence to one another as you interact, learn, and practice helping skills. Recall especially the basic elements of physical attending summarized by the acronym *SOLER*:

S   Face your clients **SQUARELY**. This says that you are available to work with them.
O   Adopt an **OPEN** posture. This says that you are open to your clients and want to be nondefensive.
L   **LEAN** toward the client at times. This underscores your attentiveness and lets clients know that you are with them.
E   Maintain good **EYE** contact without staring. This tells your clients of your interest in them and their concerns.
R   Remain relatively **RELAXED** with clients as you interact with them. This indicates your confidence in what you are doing and also helps clients relax.

Of course, these are guidelines rather than hard and fast rules. Here is a checklist to help you provide that feedback to your fellow helpers.

- How effectively is the helper using postural cues to indicate a willingness to work with the client?
- In what ways does the helper distract clients and observers from the task at hand, for instance, by fidgeting?
- How flexible is the helper when engaging in *SOLER* behaviors? To what degree do these behaviors help the counselor be with the client effectively?
- How natural is the helper in attending to the client? Are there any indications that the helper is not being himself or herself?
- From an attending point of view, what does the helper need to do to become more effectively present to his or her clients?

More important than nonverbal behavior in itself is the total quality of your being with and working with your clients. Your posture and nonverbal behavior are a part of your presence, but there is more to presence than *SOLER* activities. Since quality of presence involves both internal attitudes and external behaviors, trainees should not become preoccupied with the micro-skills of attending.

II. EXERCISES IN ACTIVE LISTENING

Read the sections on listening in Chapter Four of *The Skilled Helper*. Effective helpers are active listeners. When you listen to clients, you listen to:

- their **experiences**, what they see as happening *to* them;
- their **behaviors**, what they do or fail to do;
- their **affect**, the feelings and emotions that arise from their experiences and behaviors; and
- the **context** of both the helping interview and the daily life of the client
- the **core messages** in their stories;
- the **slant** they might give to their stories.

Helpers listen carefully in order to be able to respond with both understanding and, as we shall see later, to help clients challenge themselves.

EXERCISE 7: LISTENING TO YOUR OWN FEELINGS AND EMOTIONS

If you are to listen to the feelings and emotions of clients, you first should be familiar with your own emotional states. A number of emotional states are listed below. You are asked to describe what you feel when you feel these emotions. Describe what you feel as *concretely* as possible: How does your body react? What happens inside you? What do you feel like doing? Consider the following examples.

Example 1—Accepted:

When I feel accepted,

- I feel warm inside.
- I feel safe.
- I feel free to be myself.
- I feel like sitting back and relaxing.
- I feel I can let my guard down.
- I feel like sharing myself.
- I feel some of my fears easing away.
- I feel at home.
- I feel at peace.
- I feel my loneliness melting away.

Example 2—Scared:

When I feel scared,

- my mouth dries up.
- my bowels become loose.
- there are butterflies in my stomach.
- I feel like running away.
- I feel very uncomfortable.
- I feel the need to talk to someone.
- I turn in on myself.
- I'm unable to concentrate.
- I feel very vulnerable.
- I sometimes feel like crying.

1. Choose four of the emotions listed below or others not on the list. Try your hand at the emotions you have difficulty with. It's important to listen to yourself when you are experiencing emotions that are not easy for you to handle.
2. Picture to yourself situations in which you have actually experienced each of these emotions.
3. Then, as in the example above, write down in detail what you experienced.

1. accepted	12. disappointed	23. lonely
2. affectionate	13. free	24. loving
3. afraid	14. frustrated	25. rejected
4. angry	15. guilty	26. respected
5. anxious	16. hopeful	27. sad
6. attracted	17. hurt	28. satisfied
7. bored	18. inferior	29. shocked
8. competitive	19. interested	30. shy
9. confused	20. intimate	31. superior
10. defensive	21. jealous	32. suspicious
11. desperate	22. joyful	33. trusting

The reason for this exercise is to sensitize yourself to the wide variety of ways in which clients express and name their feelings and emotions.

LISTENING TO EXPERIENCES, BEHAVIORS, AND THE FEELINGS AND EMOTIONS THEY GENERATE

Although the feelings and emotions of clients (not to mention your own) are extremely important, sometimes helpers concentrate too much, or rather too exclusively, on them. Feelings and emotions need to be understood, both by helpers and by clients, in the **context** of the experiences and behaviors that give rise to them. On the other hand, when clients hide their feelings, both from themselves and from others, then it is necessary to listen carefully to cues indicating the existence of suppressed, ignored, or unmanaged emotion.

EXERCISE 8: LISTENING TO CORE MESSAGES

Core messages are the main points of a client's story. The ingredients of core messages are key experiences and key behaviors together with the key feelings or emotions associated with them. In this exercise you are asked to "listen to" and identify the key experiences and behaviors that give rise to the client's main feelings.

1. Listen very carefully to what the client is saying.
2. Identify the client's key experiences, what he or she says is happening to him or her.
3. Identify the client's key behaviors, what he or she is doing, not doing, or failing to do.
4. Identify the key feelings and emotions associated with these experiences and behaviors.

Example: A twenty-seven-year-old man is talking to a minister about a visit with his mother the previous day.

> I just don't know what got into me! She kept nagging me the way she always does, asking me why I don't visit her more often. As she went on, I got more and more angry. (He looks away from the counselor down toward the floor.) I finally began screaming at her. I told her to get off my case. (He puts his hands over his face.) I can't believe what I did! I called her a bitch. (Shaking his head.) I called her a bitch about ten times and then I left and slammed the door in her face.

a. **Key experiences:** Mother's nagging.
b. **Key behaviors:** Losing his temper, yelling at her, calling her a name, slamming The door in her face.
c. **Feelings/emotions generated:** He feels embarrassed, guilty, ashamed, distraught, extremely disappointed with himself, remorseful.

*Note carefully: This man is talking **about** his anger, the way he let his temper get away from him, but while talking to the minister, he is feeling and expressing the emotions listed above.*

1. A woman, 40, married, no children who has had several sessions with a counselor. She went because she was bored and felt that all the color had gone out of her life. In a later session she says this: "These counseling sessions have really done me a great deal of good! I've worked hard in these sessions, and it's paid off. I enjoy my work more. I actually look forward to meeting new people. My husband and I are talking more seriously and decently to each other. At times he's even tender toward me the way he used to be. Now that I've begun to take charge of myself more and more, there's just so much more freedom in my life!

a. **Client's key experiences:**

_____

_____

b. **Client's key behaviors:**

_____

_____

c. **How does the client feel about these experiences and behaviors?**

_____

_____

2. A man, 64, who has been told that he has terminal lung cancer, is speaking to a medical resident: "Why me? Why me? I'm not even that old! I keep looking for answers and there are none. I've sat for hours in church and I come away feeling empty. Why Me? I don't smoke or anything like that. (He begins to cry.) Look at me. I thought I had some guts. I'm just a slobbering mess. Oh God, why terminal? What are these next months going to be like? (Pause, he stops crying.) Why would you care! I'm just a failure to you guys."

a. **Client's key experiences:**

_____

_____

b. **Client's key behaviors:**

_____

_____

c. **How does the client feel about these experiences and behaviors?**

_____

_____

3. A woman, 38, unmarried, talking about a situation with a friend: "My best friend has just turned her back on me. And I don't even know why! (said with great emphasis) From the way she acted, I think she has the idea that I've been talking behind her back. I simply have not! (also said with great emphasis)

Damn! This neighborhood is full of spiteful gossips. She should know that. If she's been listening to those foulmouths who just want to stir up trouble. . . . She could at least tell me what's going on."

a. **Client's key experiences:**

_____

_____

b. **Client's key behaviors:**

_____

_____

c. **How does the client feel about these experiences and behaviors?**

_____

_____

4. A man, 54, talking to a counselor about a situation at work: "I don't know where to turn. They're asking me to do things at work that I just don't think are right. If I don't do them—well, I'll probably be let go. And I don't know where I'm going to get another job at my age in this economy. But if I do what they want me to, I think I could get into trouble, I mean legal trouble. I'd be the fall guy. My head's spinning. I've never had to face anything this before. . . . Where do I turn?"

a. **Client's key experiences:**

_____

_____

b. **Client's key behaviors:**

_____

_____

c. **How does the client feel about these experiences and behaviors?**

_____

_____

5. A girl in her late teens who is serving a two-year term in a reformatory speaks to a probation counselor: (She sits silently for a while and doesn't answer any question the counselor puts to her. Then she shakes her head and looks around the room.): "I don't know what I'm doing here. You're the third counselor they've sent me to . . . or is it the fourth? It's a waste of time! Why do they keep making me come here? (She looks straight at the counselor.) Let's fold the show right now. You're not getting anything out of me. Come on, get smart."

a. **Client's key experiences:**

_____

_____

b. **Client's key behaviors:**

_____

_____

c. **How does the client feel about these experiences and behaviors?**

_____

_____

**The following are supplemental exercises.
They should be filled in only if the you have not yet gotten the point or
feel that further written practice will help you master this part of the helping model.**

A. A man, 45, with a daughter, 14, who was hit by a car a few days earlier: "I should never have allowed my daughter to go to the movies alone. (He keeps wringing his hands.) My wife is still extremely upset. (He grimaces.) She says I'm careless . . . but being careless with the kids . . . that's something else! (He stands up and walks around.) I almost feel as if *I* had broken Karen's arm, not the guy in that car. (He sits down, stares at the floor, keeps tapping his fingers on the desk.) I don't know."

a. **Client's key experiences:**

_____

_____

b. **Client's key behaviors:**

_____

_____

c. **How does the client feel about these experiences and behaviors?**

_____

_____

B. A trainee, 29, speaking to the members of his counselor training group. It is early in the life of the group and each trainee has been asked to share thoughts about being in the group: "I don't know what to expect in this group. (He speaks hesitatingly.) I've never been in this kind of group before. From what I've seen so far, I, well, I get the feeling that you're pros, and I keep watching myself to see if I'm doing things right. (Sighs heavily.) I'm comparing myself to what everyone else is doing. I want to get good at this stuff . . . (pause) . . . but frankly I keep wondering how good I can get at this . . . or whether I can make it at all. . . . There! The cat is out of the bag."

a. **Trainee's key experiences:**

_____

_____

b. **Trainee's key behaviors:**

_____

_____

c. **How does the trainee feel about these experiences and behaviors?**

_____

_____

C. A woman, 42, married, with three children in their early teens, is speaking to a church counselor: "Why does my husband keep blaming me for his trouble with the kids? I'm always in the middle. He complains to me about them. They complain to me about him. (She looks the counselor straight in the eye and talks very deliberately.) I could walk out on the whole thing right now. Who the hell do they think they are?"

a. **Client's key experiences:**

_____

_____

b. **Client's key behaviors:**

_____

c. How does the client feel about these experiences and behaviors?

_____

_____

D. A bachelor, 39, speaking to the members of a group in which the members meet bi-weekly to discuss important issues in their lives. They call it a "lifestyle" group. He has belonged to this group for about a year: "I've talked a lot about my rather meager social life. But I've finally met a woman who is very genuine and who lets me be myself. We seem to be able to care deeply about each other without feeling dependent on each other. (He is speaking in a soft, steady voice.) I've always had a horror of dependency—going either direction. And she cares about me without mothering me. I never thought it would happen. (He raises his voice a bit.) Is this actually happening to me? Is it actually happening?"

a. **Client's key experiences:**

_____

_____

b. **Client's key behaviors:**

_____

_____

c. **How does the client feel about these experiences and behaviors?**

_____

_____

EXERCISE 9: THE SHADOW SIDE OF LISTENING

Since effective helpers are active listeners, it helps to learn about your own listening style—both the upside and the downside.

1. Do this part of the exercise before reading **THE SHADOW SIDE OF LISTENING TO CLIENTS** section of Chapter Four. What kind of listener are you? Give a brief description.

_____

_____

_____

_____

_____

_____

2. Next read **THE SHADOW SIDE OF LISTENING TO CLIENTS.** Then indicate below any areas of listening you think you should work on.

_____

_____

_____

_____

_____

_____

Discuss your description of yourself as a listener together with your shadow-side assessment with a learning partner. Talk about implications and possible actions you should take to become a skilled active listener.

Section 5
BASIC EMPATHY

The payoff of attending and listening lies in the ability to communicate to clients an understanding of their experiences and behaviors and the feelings and emotions they generate. Furthermore, listening to clients' points of view enables you to let them know that you see their point of view even when you think that this point of view needs to be challenged or transcended. There are two parts to this section. Part I focuses on empathy; Part II adds the skill of probing.

COMMUNICATING UNDERSTANDING: EXERCISES IN BASIC EMPATHY

Basic empathy is the communication to another person of your understanding of his or her point of view with respect to his or her experiences, behaviors, and feelings. It is a skill you need throughout the helping model. Focusing on the client's point of view without necessarily agreeing with it is very useful in establishing and developing relationships with clients and in helping them clarify both problem situations and unexploited opportunities, in setting goals, and in developing strategies and plans. The starting point of the entire helping process and each of its steps is the client's point of view, even when it needs to be challenged.

The exercises in the previous section emphasized your ability to listen to and understand the client's point of view. The exercises in this section relate to your ability to **communicate** this understanding to the client.

EXERCISE 10: IDENTIFYING CORE MESSAGES

In this exercise, you are asked to identify key experiences, behaviors, and feelings and then translate them into a core message.

1. Listen carefully to the client's statement.
2. Identify key experiences, behaviors, and the feelings they generate.
3. Pull them together in a statement of the client's core message.

Example: Roula is recovering from a bad car accident in a rehabilitation unit. She is a week or two into a program that could well go on for a few months. She is talking with one of the rehabilitation specialists.

> You told me it would be tough going. And I thought that I had prepared myself for it. I thought I had some courage in me. But now I can't find any grit at all. Doing the smallest things takes so much effort! I keep breaking down and crying when I'm alone. I just keep giving in. . . . I'm just inches down a path that seems miles long. It seems endless. . . . I can't. . . . I just can't. . . . I. . . .

Key experience(s): finding the rehabilitation program so difficult

Key behavior(s): inability to find her courage; giving in to discouragement

Key feelings/emotions: disappointment, discouragement, despair

Core message: Roula seems disappointed almost to the point of despair because she has failed to tap into the "stuff" within her that will keep her going on a very tough rehab program.

1. This woman is about to go to her daughter's graduation from college. She is talking with one of the ministers of her church: "I never thought that my daughter would make it through. I've invested a lot of money in her education. It meant scrimping and saving and not having some of the things I wanted. But money is certainly not the issue. More to the point, I've put a lot of emotion into making this day happen. I had to do a lot of hand-holding to help her get through. There were times when neither of us thought this day would come. But the day has arrived!"

Key experience(s): _____

_____

Key behavior(s): _____

_____

Key feelings/emotions: _____

Core message: _____

_____

_____

2. This older man has just had his wallet stolen. He became disoriented and was taken to a hospital. He talking with a social worker. "I had just cashed my paycheck and the money was in the wallet. I've had a streak of bad luck. My sister was in an auto accident last week. And I was on my way to visit my nephew. Ironically, he was detained by the police for a minor theft earlier this week. Now he's probably thinking I've given up on him, when I haven't. There has not been much good news at all or a while."

Key experience(s): _____

_____

Key behavior(s): _____

_____

Key feelings/emotions: _____

_____

Core message: _____

_____

_____

3. This man is waiting for the results of medical tests. He is talking to a hospital volunteer: "I've been losing weight for about two months and feeling tired and listless all the time. I'm afraid of what these tests are going to say. I know I've been putting them off. I just hate hospitals and this kind of stuff. Well, now the waiting's getting to me. I . . . well, I just don't know where I stand. Nobody said anything to me during the tests. I don't think that's a good sign. It's all so impersonal anyway."

Key experience(s): _____

_____

Key behavior(s): _____

_____

Key feelings/emotions: _____

e message: _____

_____

4. A prospective employer has just found out that this client has a criminal record: "I had hoped that I would get the job and prove myself before anyone found out about my record. I guess I was just stupid. I've just received a call from him telling me that I'm no longer being considered for the job. Well, I did what I thought was right. I never had the intention of deceiving anyone. I mean that I didn't think that I was doing anything wrong. . . . Well, even though I'm branded because of my record, I'm going to make it, somehow."

Key experience(s): _____

_____

Key behavior(s): _____

_____

Key feelings/emotions: _____

_____

Core message: _____

_____

_____

5. This woman has just lost a custody case for her only son. During the interview with a counselor she seems almost in a daze: "I never dreamed that the court would award custody to my husband. I may not be a perfect mother, but I've done all the work in raising him so far. He's done nothing but give me a hard time over the years. He's so selfish and spiteful. I've been racking my brains trying to come up with something I can do. . . . Now he's won. . . . It's all over."

Key experience(s): _____

_____

Key behavior(s): _____

_____

Key feelings/emotions: _____

27

Core message: _____

_____

_____

4. After finishing, compare your responses with those of your learning partner. Then see if the two of you can improve upon your combined responses.

EXERCISE 11: USING A FORMULA TO EXPRESS EMPATHY

Empathy focuses on the client's key messages—key experiences and/or key behaviors plus the feelings and emotions they generate. In this exercise you are asked to use the formula explained in the text: "You feel . . ." (followed by the right emotion and some indication of its intensity) "because . . ." (followed by the key experiences and/or behaviors that give rise to the emotion).

1. Identify for yourself the speaker's key experiences, behaviors, and feelings.
2. Formulate an empathic response. Even though you are writing the response, picture yourself actually talking to the person.

Example: A woman in a self-help group is talking about a relationship with a man. She says:

> About a couple of months ago, he began being abusive, calling me names, describing my defects. To tell you the truth, that's why I joined this group, but I haven't had the courage to talk about it till now. The couple times I've tried to stand up for myself, he became even more abusive. He hasn't hit me or anything, but. . . . So I've been just taking it, just sitting there taking it . . . like a dog or something. Do you think that this is just his bizarre way of getting rid of me? Why doesn't he just tell me?

Key experience(s): being abused by companion, escalation in abuse

Key behavior(s): trying to stand up for herself, becoming passive, trying to figure out if this is his way of getting rid of her

Key feelings/emotions: distraught, confused, angry

Empathic response (using formula): You feel angry and confused because the abuse came from out of the blue and now you're wondering whether this is just his strange way of ending the relationship.

1. First-year college student talking to a counselor in the Center for Student Services. She has been talking about some of the difficulties of adjusting to college life. She comes from a small town and is attending a large state university. She speaks openly and seems to be in good spirits; she even smiles at times: "My friends seem so much more sophisticated than I am. Just the other night Mairead and Johanna showed up at the party in dresses _and_ with dates. And there I was, alone and in jeans! I mean, just typical of me!

. . . I guess I was more amazed than embarrassed. And I did have a good time anyway. It's like I have to learn a whole new set of rules. . . . But nobody just comes out and says what they are. . . . Well, they don't do that anywhere, do they?"

Key experience(s): _____

_____

Key behavior(s): _____

_____

Key feelings/emotions: _____

_____

Empathy (formula): _____

_____

2. A 66-year-old man is talking to a mental health counselor: "My wife died last year, and this year my youngest son went away to college. The other children are married. So now that I'm retired, I spend a lot of time rambling around a house that's . . . . (pauses, look out the window for a while) . . . really too big for me. . . . You know, when I was working, there was a certain fullness to life. I always knew what to do. I, well . . . I made a difference. Now that I've got a comfortable retirement, I . . . ."

Key experience(s): _____

_____

Key behavior(s): _____

_____

Key feelings/emotions: _____

_____

Empathy (formula): _____

_____

3. This 33-year-old woman has been working with a counselor to find ways of handling sex discrimination at work. The company seem to preach one philosophy but implement another. Up to this point she has had some modest success, but now she wants to discuss a setback: "It seems they let me push as long as I stayed with the little stuff and did it quietly. But last week I brought up sex discrimination in our monthly team meeting . . . and my boss clammed up. And he's not one of the worst

offenders. He's been quite cool since then and always seems to be busy. All of a sudden there's quite a different atmosphere in the office."

Key experience(s): _____

_____

Key behavior(s): _____

_____

Key feelings/emotions: _____

_____

Empathy (formula): _____

_____

4. Man, 40, talking about his invalid mother to a minister at his church: "She uses her illness to control me. It's a pattern; she's been controlling me all her life. The times I try to stand up to her, well, she has her way of dealing with that, too. (He grits his teeth and sets his jaw.) I bet she'll even make me feel guilty for her death."

Key experience(s): _____

_____

Key behavior(s): _____

_____

Key feelings/emotions: _____

_____

Empathy (formula): _____

_____

5. Young woman, 25, talking about her current boyfriend to an older confidante: "I can't quite figure him out. I still can't tell if he really cares about me, or if he's just trying to get me into bed. It leaves everything up in the air. He's pleasant enough, but I'm not finding the substance that I thought was there. I do try to talk about serious things, things that interest me like what's happening in the world. The next thing I know we're back into trivia. I've been tempted to talk about this directly, but something holds me back."

Key experience(s): _____

_____

Key behavior(s): _____

_____

Key feelings/emotions: _____

_____

Empathy (formula): _____

_____

6. This young man has just been abandoned by his wife after only a year of marriage: "We've been married for about a year. . . . She left a note saying that this has not been working out. Just like that. You know, I thought that things were going fairly well. Not perfect, of course. We had our ups and downs, but everyone does. Maybe I was so busy at work that there were things I didn't notice. I don't know if there's someone else. I really don't know why. . . . I must really sound stupid. I have no idea what to do to get her back. You can't make someone love you"

Key experience(s): _____

_____

Key behavior(s): _____

_____

Key feelings/emotions: _____

_____

Empathy (formula): _____

_____

7. This woman has been suffering from migraine headaches for a long time: "They seem to be getting worse. So far nothing has helped me reduce their number or to manage them once they start. I won't go to a doctor. They've never helped. Oh, I read about those so-called 'new' treatments, but I bet it's the same old stuff promising you a lot and delivering nothing."

Key experience(s): _____

Key behavior(s): _____

_____

Key feelings/emotions: _____

_____

Empathy (formula): _____

_____

8. This man is talking about having to work two jobs to support his family: "I guess I'm fortunate to have both jobs, but I've got no time for myself. The jobs eat into my evening hours and the weekends. I think what really bothers me is that my family does not seem to notice my absence. They take for granted all the hours I've been putting in. It's as if life is about nothing else but work and no one else cares."

Key experience(s): _____

_____

Key behavior(s): _____

_____

Key feelings/emotions: _____

_____

Empathy (formula): _____

_____

9. The following person, Alfred, is being forced out of the company where he has worked for almost twenty years because the company is downsizing: "It came out of nowhere. I came in two days ago and there on my desk was a notice saying that I was terminated. It was as impersonal as that. . . . I kept reading it over and over again like I was in some dream. I tried to call my supervisor but couldn't get her. Later I found out that she had been let go, too. The notice said I get some *outplacement* help. What kind of word is that! It's a fancy word for dumping."

Key experience(s): _____

_____

Key behavior(s): _____

Key feelings/emotions: _____

Empathy (formula): _____

10. The following trainee is talking to her instructor. The training group has just been introduced to the skill of challenging clients. They have practiced self-challenge and have just begun using challenge with one another: "For the first time in this program I having trouble and it's bothering me. What I've found out is that I have no problem with self-challenge. In some ways I have been doing that for years. But now I see that maybe I've been picking on myself, you know, being overly perfectionistic. But now I'm finding it really hard to challenge the members of my group. Oh, I know the theory. But doing it is hard. It's almost as if they won't like me if I do it well."

Key experience(s): _____

Key behavior(s): _____

Key feelings/emotions: _____

Empathy (formula): _____

EXERCISE 12: USING YOUR OWN WORDS TO EXPRESS EMPATHY

In this exercise you are asked to do two things:

1. First use the "you feel . . . because . . ." formula to communicate empathy to the client.
2. Then recast your response in your own words while still identifying both core message(s) in terms of key experiences and/or behaviors together with the feelings that they generate.

Example: A married woman, 31, is having a "solo" meeting with a marriage counselor to review how things have been going: "I can't believe it! You know when Tom and I were here last week we made a contract that said that he would be home for supper every evening and on time. Well, he came home on time every day this past week. I never dreamed that he would live up to his part of the bargain so completely! I've even begun making better meals!"

• **Formula.** "You feel great because he really stuck to the contract!"
• **Non-formula.** "Making good on his word has given you quite a boost!"

Now imagine yourself listening intently to each of the clients quoted below. First use the "You feel . . . because . . ." formula; then use your own words. Try to make the second response sound as natural (as much like yourself) as possible. After you use your own words, check to see if you have both a "you feel" part and a "because" part in your response.

1. Woman, 28, talking about her job with a colleague: "It's not a big thing. But this is the third time this month I've been asked to change hours with her. I said yes, of course. But it certainly seems to indicate who is more important there. Why does it always have to be me who defers to her?"

a. Use the formula.

_____

_____

b. Use your own words.

_____

_____

2. A man who has been suffering from depression for a number of months is talking with a psychologist: "I've seen my doctor a couple of times. But now since the depression isn't getting any better, he wants to give me this drug. . . . But I don't even take aspirin if I don't have to. Somehow drugs and me don't mix. If I had pneumonia or some infection, it would be one thing. But this is all in my head. I don't want to be seen as a psycho!"

a. Use the formula.

_____

_____

b. Use your own words.

_____

_____

3. A young women, 20, is speaking to a college counselor toward the end of her second year: "I've been in college almost two years now, and nothing much has happened. (She speaks listlessly.) The teachers here are only so-so. I thought they'd be a lot better. At least that's what I heard. And I can't say much for the social life here. Things go on the same from day to day, from week to week. To be honest, it's a rather boring place. . . . I keep thinking about this more and more."

a. Use the formula.

_____

_____

b. Use your own words.

_____

_____

4. A fellow trainee has been talking with you about the counselor training program. He call you up and says he wants to talk with you right away. When you meet, he says: "You know, tomorrow we're going to start talking about our own problems. Well, the kinds of things that could stand in the way of being good helpers. . . . Well, there are things I'd rather not say in the group. . . . I even had to screw up my courage to talk with you. . . . You know, a couple of issues I'd rather not talk about *could* stand in the way of my being the kind of counselor I should be. . . . But I'm just not ready. I could be dishonest."

a. Use the formula.

_____

_____

b. Use your own words.

_____

_____

5. A businessman, 38, talking to a company counselor: "I really don't know what my boss wants. I don't know what he thinks of me. He tells me I'm doing fine even though I don't think that I'm doing anything special. Then he blows up over nothing at all. I keep asking myself if there's something wrong with me, I mean, that I don't see what's getting him to act the way he does. I'm beginning to wonder if this is the right job for me."

a. Use the formula.

_____

_____

b. Use your own words.

_____

6. A seventh-grade girl to her teacher, outside class: "My classmates don't like me, and right now I don't like them! Why do they have to be so mean? They make fun of me—well, they make fun of my clothes. My family can't afford what some of those dopes wear. Gee, they don't have to like me, but I wish they'd stop making fun of me."

a. Use the formula.

_____

_____

b. Use your own words.

_____

_____

The following are supplemental exercises.
They should be filled in only if the you have not yet gotten the point or
feel that further written practice will help you master this part of the helping model.

A. Student, 16, talking to a teacher whom he trusts about his a teacher with whom he has not been getting on: "I thought he was going to really chew me out. I was afraid that he was just going to tell me he was kicking me out of class. But we sat in his office and he said that we should talk about our differences. And we did!"

a. Use the formula.

_____

_____

b. Use your own words.

_____

_____

B. This woman has been suffering from migraine headaches for a long time: "They seem to be getting worse. So far nothing has helped me reduce their number or to manage them once they start. I won't go to a doctor. They've never helped. Oh, I read about those so-called 'new' treatments, but I bet it's the same old stuff promising you a lot and delivering nothing."

a. Use the formula.

_____

b. Use your own words.

_____

_____

C. This man is talking about having to work two jobs to support his family: "I guess I'm fortunate to have both jobs, but I've got no time for myself. The jobs eat into my evening hours and the weekends. I think what really bothers me is that my family does not seem to notice my absence. They take for granted all the hours I've been putting in. It's as if life is about nothing else but work and no one else cares. To make things worse, every now and again they complain when I'm not around when they say they need me."

a. Use the formula.

_____

_____

b. Use your own words.

_____

_____

D. A woman, 73, in the hospital with a broken hip, is talking with a pastoral care volunteer: "When you get old, you have to expect things like this to happen. It could have been much worse. When I lie here, I keep thinking of the people in the world who are a lot worse off than I am. I'm not a complainer. Oh, I'm not saying that this is fun or that the people in this place give you the best service—who does these days?—but it's a good thing that these hospitals exist. Think of those who don't have anything."

a. Use the formula.

_____

_____

b. Use your own words.

_____

_____

E. Man, 35, talking to a counselor at the company where he works (he wrings his hands as he talks): "I'm going to the hospital tomorrow for some tests. The doctor suspects an ulcer. But nobody has told me exactly what the tests are supposed to show. . . . I've got this prep kit from the drug store. And I'm not

37

going to seat anything after supper this evening. . . . I've heard rumors about what these tests are like but I don't really know. I've never gone through this before. . . . I'm pretty young for all this."

a. Use the formula.

_____

_____

b. Use your own words.

_____

_____

F. A woman, 53, about to get divorced, is talking with a counselor: "My husband and I have just decided to get a divorce. (Her voice is very soft, her speech is slow, halting.) I think it's more like he decided. . . . I really don't look forward to the legal part of it (pause) to *any* part of it to tell the truth. For the first time in my life I just sit around and think a lot. Or stare into space. I just don't know what to expect. (She sighs heavily.) I'm well into middle age. I don't think another marriage is possible. I just don't know what to expect."

a. Use the formula.

_____

_____

b. Use your own words.

_____

_____

EXERCISE 13: EMPATHY WITH CLIENTS FACING DILEMMAS

Clients sometimes talk about complicated issues. When this is the case, it is essential to listen even more carefully in order to pick up and respond to core messages. For example, clients often talk about conflicting values, experiences, behaviors, and emotions. Responding with empathy means communicating an understanding of the conflict.

Example: A woman, 32, talking to a counselor about adopting a child:

> I'm going back and forth, back and forth. I say to myself, 'I really want a child,' but then I think about Bill [her husband] and his reluctance. He so wants our own child and is so reluctant to raise someone else's. We don't even know why we can't have children. But the fertility specialists don't offer us much hope. At times when I so want to be a mother I think I should marry someone who

would be willing to adopt a child. But I love Bill and don't want to point an accusing finger at him.

- **Identify the conflict or dilemma.** She believes that she runs the risk of alienating her husband if she insists on adopting a child, even though she strongly favors adoption.
- **Formula.** "You feel trapped between your desire to be a mother and your love for your husband."
- **Non-formula.** "You're caught in the middle. Adopting a child would solve one problem but perhaps create another."

1. A factory worker, 30: "Work is okay. I do make a good living, and both my family and I like the money. My wife and I are both from poor homes, and we're living much better than we did when we were growing up. But the work I do is the same thing day after day. I may not be the world's brightest person, but there's a lot more to me than I use on those machines."

a. The conflict.

_____

b. Use the formula.

_____

_____

c. Use your own words.

_____

_____

_____

2. A mental hospital patient, 54, who has spent five years in the hospital; he is talking to the members of an ongoing therapy group. Some of the members have been asking him what he's doing to get out. He says, "I don't know why you're trying to push me out of here. To tell the truth, I like it here. I'm safe and secure. So why are so many people here so damn eager to see me out? . . . Is it a crime because I feel comfortable here? (Pause, then in a more conciliatory voice.) I know you're all interested in me. I see that you care. But do I have to please you by doing something I don't want to do?"

a. The conflict.

_____

b. Use the formula.

_____

_____

c. Use your own words.

_____

_____

_____

3. A juvenile probation officer to a colleague: "These kids drive me up the wall. Sometimes I think I'm really stupid for doing this kind of work. They taunt me. They push me as far as they can. To some of them I'm just another 'pig.' But every time I think of quitting—and this gets me—I know I'd miss the work and even miss the kids one way or another. When I wake up in the morning, I know the day's going to be full and it's going to demand everything I've got."

a. The conflict.

_____

b. Use the formula.

_____

_____

c. Use your own words.

_____

_____

_____

EXERCISE 14: THE PRACTICE OF EMPATHY IN EVERYDAY LIFE

If the communication of accurate empathy is to become a part of your natural communication style, you will have to practice it outside formal training sessions. That is, it must become part of your everyday communication style or it will tend to lack genuineness in helping situations. Practicing empathy "out there" is a relatively simple process.

1. **Empathy as an improbable event.** Empathy is not a normative response in everyday conversations. Find this out for yourself. Observe everyday conversations. Count how many times empathy is used as a response in any given conversation.

2. **Your own use of empathy.** Next try to observe how often you yourself use empathy as part of your normal style. In the beginning, don't necessarily try to increase the number of times you use empathy in day-to-day conversations. Merely observe your usual behavior. What part does empathy normally play in your style?

3. **Increasing your empathic responses.** Begin to increase the number of times you use accurate empathy. Be as natural as possible. Do not overwhelm others with this response; rather try to incorporate it gradually into your style. You will probably discover that there are quite a few opportunities for using empathy without being phony. Find some way of keeping track of your progress.

4. **The impact of empathy.** Observe the impact your use of empathy has on others. Don't set our to use others for the purpose of experimentation but, as you gradually increase your use of this communication skill naturally, try to see how it influences your conversations. What impact does it have on you? What impact does it have on others?

5. **Learnings.** In a forum set up by the instructor, discuss with your fellow trainees what you are learning from this experiment.

If empathy becomes part of your communication style "out there," then you should appear more and more natural in using empathy in the training program, both in playing the role of the helper and in listening and providing feedback to your fellow trainees. On the other hand, if you use empathy only in the training sessions, it will most likely remain artificial.

Section 6
THE ART OF PROBING AND SUMMARIZING

Problems and opportunities are better managed when they are clear and specific. If Connie and Chuck want to deal with the poor communication they have with each other in their marriage, then the key issues relating to communication need to be spelled out in terms of specific experiences, specific behaviors, and specific feelings in specific situations. Clients need to focus on concrete rather than abstract issues. Counselors use probes to help clients "fill in the picture," turning vague descriptions of experiences, behaviors, and feelings into concrete and specific ones. Vaguely described problems lead only to vague solutions. Read Chapter Six in *The Skilled Helper* before doing these exercises. This section has three parts: first, problem and opportunity clarity and specificity; second, probing; third, summarizing.

I. PROBLEM AND OPPORTUNITY CLARITY

EXERCISE 15: SPEAKING CONCRETELY ABOUT PROBLEMS

This exercise is designed to help you get a feel for both vagueness and concreteness. In this exercise you are asked to focus on your own story in terms of your own experiences, behaviors, and feelings.

1. Give a vague statement of some problem situation.
2. Turn the vague statement into a concrete and specific statement in terms of key experiences, behaviors, and feelings.

As usual, you are asked to focus on issues—either problems or unused opportunities—that can make a difference to you in your role as helper.

Example 1: Karen writes about her need to control situations.

- **Vague statement of problem situation**: "I tend to be a bit domineering at times."

- **Concrete statement**: "I try, usually in subtle ways, to get others to do what *I* want to do. I even pride myself on this. In conversations, I take the lead. I interrupt others, jokingly and in a good-natured way, but I make my points. If a friend is talking about something serious when I'm not in the mood to hear it, I change the subject."

Example 2: Jamie talking about how the training group affects him.

- **Vague statement problem situation**: "I get bothered in training groups."
- **Concrete statement**: "I feel hesitant and embarrassed whenever I want to give feedback to other trainees, especially if it is in any way negative. When the time comes, my heart beats faster and my palms sweat. I feel like everyone is staring at me, though I know they're not. I feel under pressure here. Outside I have a live-and-let-live mentality. I'm not used to giving feedback to anyone."

In the spaces below explore three issues that are related to some problem situation or situations of your own. Try to choose issues that might affect the quality of your helping.

1a. Vague.

_____

_____

1b. Concrete.

_____

_____

_____

_____

_____

_____

2a. Vague.

_____

_____

2b. Concrete.

_____

_____

_____

_____

_____

_____

_____

Share one or two of these with a learning partner. Get feedback on how clear your statement is. If you do not think that your partner's statement is as clear as it might be, use probes to help him or her make the statement clearer.

EXERCISE 16: SPEAKING CONCRETELY ABOUT UNUSED OPPORTUNITIES

In this exercise, you are asked to speak about some of your unused or underused resources or opportunities. As in the exercise above, start with a vague statement, then clarify it with the kind of detail needed to serve the opportunity-development process. Try to choose unused resources or opportunities that might add value to your role as helper.

1. Give a vague statement of some unused or underused resource or opportunity.
2. Turn the vague statement into a concrete and specific statement in terms of key experiences, behaviors, and feelings.

Example 1: Jane, a counselor trainee, discusses her communication style.

- **Vague statement of unused opportunity**: "There is a very good communicator inside me that doesn't get out enough."
- **Concrete statement**: "This course has shown me that I am already fairly good at many of the communication skills we are learning. I listen well. I'm empathic. From what I understand about challenge, I think that I can even challenge people without turning them off. But so often I choose to be passive. It groups I'm often an onlooker rather than a participant. I found out that I can't get away with that in this group. I think I need to become much more assertive in order to be an effective learner and helper."

Example 2: Austin, in his fourth year of a doctoral program in clinical psychology, has also taken a couple of courses in organizational psychology. He has just finished his internship.

- **Vague statement of possible opportunity**: "One thing I learned during my internship rotations is that mental-health organizations are not always well run. But I'm not sure I can do anything about it without violating my commitment to people."
- **Concrete statement**: "The organizational psychology courses I took made it clear to me that some mental-health centers are poorly-run organizations. Even with a little background I knew some things I could do to help them be more efficient and effective. In fact, the whole area of applying effective management and organizational ideas to mental-health systems appeals to me enormously. But I feel

guilty about pursuing that possibility after graduation. It's like betraying the reasons for going into clinical psych in the first place—mainly to help others. I really liked dealing with individuals and groups during my internship, but I was distracted by how poorly these places are administered. A number of psychologists I met thought that administration was for the birds, you know, for lesser people."

In the spaces below, deal with three instances of unused or underused resources or opportunities. Choose situations that are could add value to your role as helper.

1a. Vague.

_____

_____

1b. Concrete.

_____

_____

_____

_____

_____

_____

2a. Vague.

_____

_____

2b. Concrete.

_____

_____

_____

_____

Share one or both of these with the your learning partner. Get feedback on how clear your second statement is. If you do not think that someone's statement is as clear as it might be, use probes to help your fellow trainee make his or her statement clearer.

II. PROBES

EXERCISE 17: PROBING FOR KEY ISSUES AND CLARITY

A probe is a statement or a question that invites a client to discuss an issue more fully. In the previous exercises you were asked to probe yourself. Probes are ways of getting at important details that clients do not think of or are reluctant to talk about. They can be used at any point in the helping process to clarify issues, search for missing data, expand perspectives, and point toward possible client actions. An overuse of probes can lead to gathering a great deal of irrelevant information. The purpose of a probe is not information for its own sake, but data—experiences, behaviors, and feelings—that serve the process of problem management and opportunity development.

In this exercise, brief problem situations together with a bit of context will be presented. Your job is to formulate possible probes.

1. First respond with empathy.
2. Formulate a probe that might help the client further identify or explore key issues in terms of experiences, behaviors, and/or feelings.
3. Briefly state your reasons for using the probe.

Example: A man, 24, complains that he is severely tempted to go on experimenting sexually with women other than his wife:

> Although I have not had an extended affair, I have had a few sexual encounters and feel that some day I might have an affair. I don't blame myself or my wife, though I'm not sure I'll ever find the kind of sexual satisfaction I want with her. I've never had such strong sexual urges. I guess I'm not thinking straight. What's going to happen to my marriage? In fact, where does this all lead? Have I just reverted to adolescence? I'm asking myself all sorts of questions.

a. **Empathic statement**: "It sounds like your sexual urges are so strong that right now that they're in the driver's seat. But you're puzzled and are not quite sure what all this is all about and where it is going."
b. **Possible probe**: "When you ask yourself questions like 'What's going to happen to my marriage," what are some of the answers you come up with?"
c. **Reason for probe**: To find out how far he has thought through some of the implications of his current thinking and behaving.

1. Grace, 19, an unmarried, first-year college student, comes to counseling because of an unexpected and unwanted pregnancy: "Right now I realize that the father could be either of two guys. That probably says

45

something about me right there. I'm not sure what I want to do about the baby. I haven't told my parents yet, but I think that they will be very upset, but in the end sympathetic. But I've gotten myself into this mess and I have to get myself out."

a. First, give an empathic response.

_____

_____

b. A probe to help client get at key issues more clearly and fully.

_____

_____

c. The reason for the probe.

_____

_____

2. You are a counselor in a halfway house. You are dealing with Tom, 44, who has just been released from prison where he served two years for armed robbery, his first encounter with the law. He has been living at the halfway house for two weeks. That is the only offense for which he has ever been convicted. The halfway house experience is designed to help him reintegrate himself into society. Living in the house is voluntary. The immediate problem is that Tom came in drunk a couple of nights ago. He was supposed to be out on a job-search day. Drinking is against the rules of the house. When you talk to him, he says "I know this looks bad. . . . But I don't think it's as bad as it looks. There was no liquor inside. I went overboard. . . . I guess I'm still a bit confused."

a. First, give an empathic response.

_____

_____

b. A probe that might help the client get at key issues more clearly and fully.

_____

_____

c. The reason for the probe.

_____

_____

3. Arnie is a born-again Christian. He has begun to do a fair amount of informal preaching at his place of employment. While some of his co-workers sympathize with his views, others are turned off. Since he feels that he is being driven by a "clear vision," he becomes more and more militant. His supervisor has cautioned him a couple of times, but this has done little to change Arnie's behavior. Finally, he is given an ultimatum to talk to one of the counselor's in the Employee Assistance Program about these issues or be suspended from his job. He says to the counselor, "I have a duty to spread the word. And if I have a duty to do so, then I also have the right. I'm a good worker. In fact, I believe in hard work. So it's not like I'm taking time off for the Lord's work. Now what's wrong with that?"

a. First, give an empathic response.

_____

_____

b. A probe that might help the client get at key issues more clearly and fully.

_____

_____

c. The reason for the probe.

_____

_____

4. A trainee in the counselor training program makes this statement in his small group: "I do not take criticism well. When I receive almost any kind of negative feedback, I usually smile and seem to shrug it off, but inside I begin to pout. Also, deep inside, I put the person who gave me the feedback on a 'list.' Those on the list have to pay for what they did. I find this hard to admit, even to myself. It sounds so petty. For instance, two weeks ago in the training group I received some negative feedback from Cindy. I felt angry and hurt because I thought she was my friend. Since then I've tried to see what mistakes she makes here. I've even felt a bit disappointed because I haven't been able to catch her in any kind of glaring mistake. It goes without saying that I'm embarrassed to say all this."

a. First, give an empathic response.

_____

_____

b. A probe that might help the client get at key issues more clearly and fully.

_____

_____

c. The reason for the probe.

_____

_____

5. Renata is talking with a counselor about her relationship with her mother: "I feel guilty and depressed whenever my mother calls and implies that she's lonely. I then get angry with myself for giving in to guilt so easily. Then the whole day has a pall over it. I get nervous and irritable and take it out on others. Or I brood. Sometimes it even interferes with my work. Then I do what I have to."

a. First, give an empathic response.

_____

_____

b. A probe that might help the client get at key issues more clearly and fully.

_____

_____

c. The reason for the probe.

_____

_____

EXERCISE 18: COMBINING EMPATHY WITH PROBES

This exercise asks you to combine several skills—the ability to be empathic, to identify areas needing clarification, and to use probes to make clients aware of the need for action. Remember that each step of the helping process should be, all things considered, some kind of stimulus for client action, the "little" actions, as it were, that precede formal action based on a plan.

1. First reply to the client with empathy.
2. Identify an area needing exploration and clarification.
3. Use a probe in open-ended question form to help the client explore or clarify some issue.
4. Convert the probe into a statement.

Example 1: A law student, 25, is talking to a school counselor:

> I learned yesterday that I've flunked out of school and that there's no recourse.
> I've seen everybody, but the door is shut tight. What a mess! I know I haven't
> gotten down to business the way I should. This is my first year in a large city
> and there are so many distractions. And school is so competitive. I have no idea

how I'll face my parents. They've paid for my college education and this year of law school. And now I'll have to tell them that it's all down the drain."

- **Empathy:** "The whole situation sounds pretty desperate both here and at home. And it sounds so final.
- **An area for probing:** It is not clear whom the client saw and what specific responses he received.
- **Question probe:** "What people did you see and which doors were shut?"
- **Statement probe:** "I'm not sure who you mean by 'everybody' and it's not clear what doors were shut."

1. A high school senior to a school counselor: "My dad told me the other night that I looked relaxed. Well, that's a joke. I don't feel relaxed. There's a lull right now, because of semester break, but next semester I'm signed up for two math courses, and math really rips me up. But I need it for science since I want to go into pre-med."

a. Empathy.

_____

_____

b. Fruitful area for probing.

_____

_____

c. Question probe.

_____

_____

d. Statement probe.

_____

_____

2. A woman, 27, talking to a counselor about a relationship that has just ended (she speaks in a rather matter-of-fact voice): "About three weeks ago I came back from visiting my parents who live in Nevada and found a letter from my friend Gary. He said that he still loves me but that I'm just not the person for him. In the letter he thanked me for all the good times we had together these last three years. He asked me not to try to contact him because this would only make it more difficult for both of us. End of story. I guess I've let my world collapse. People at work have begun complaining about me. And I've been like a zombie most of the time."

a. Empathy.

_____

_____

b. Fruitful area for probing.

_____

_____

c. Question probe.

_____

_____

d. Statement probe.

_____

_____

3. A married man, 25, talking to a counselor about trouble with his mother-in-law: "The way I see it, she is really trying to destroy our marriage. She's so conniving. And she's very clever. It's hard to catch her in what she's doing. You know, it's rather subtle. Well, I've had it! If she's trying to destroy our marriage, she's getting pretty close to achieving her goal."

a. Empathy.

_____

_____

b. Fruitful area for probing.

_____

_____

c. Question probe.

_____

_____

d. Statement probe.

_____

_____

4. A woman, 31, talking to an older woman friend: "I just can't stand my job any more! My boss is so unreasonable. He makes all sorts of silly demands on me. The other women in the office are so stuffy, you can't even talk to them. The men are either very blah or after you all the time, you know, on the make. The pay is good, but I don't think it makes up for all the rest. It's been going on like this for almost two years."

a. Empathy.

_____

_____

b. Fruitful area for probing.

_____

_____

c. Question probe.

_____

_____

d. Statement probe.

_____

_____

5. A man, 45, who has lost his wife and home in a tornado, has been talking about his loss to a social worker: "This happened to a friend of mine in Kansas about ten years ago. He never recovered from it. His life just disintegrated and nobody could do anything about it. . . . It was like the end of the world for him. You never think it's going to happen to you. I know I belong here. But this kind of thing makes me think I don't."

a. Empathy.

_____

_____

b. Fruitful area for probing.

_____

_____

c. Question probe.

_____

_____

d. Statement probe.

_____

_____

The following are supplemental exercises.
They should be filled in only if the you have not yet gotten the point or
feel that further written practice will help you master this part of the helping model.

A. A divorced woman, 44, talking to a counselor about her drinking. This is the second session. She has spent a lot of time telling her story. To the counselor there seemed to be a lot of evasions and some outright lying: "Actually, it's a relief to tell someone. I don't have to give you any excuses or make the story sound right. I drink because I like to drink; I'm just crazy about the stuff, that's all. But I'm under no delusions that telling you is going to solve anything. When I get out of here, I know I'm going straight to a bar and drink. Some new bar, new faces, some place they don't know me."

a. Empathy.

_____

_____

b. Fruitful area for probing.

_____

_____

c. Question probe.

_____

_____

d. Statement probe.

_____

_____

B. A man, 57, talking to a counselor about a family problem: "My younger brother, he's 53, has always been a kind of bum. He's always poaching off the rest of the family. Last week my unmarried sister told me that she'd given him some money for a 'business deal.' Business deal, my foot! I'd like to get hold of him and give him a good kick! Oh, he's not a vicious guy. Just weak. He's never been able to get a fix on life. But he's got the whole family in turmoil now, and we can't keep going through hell for him."

a. Empathy.

_____

_____

b. Fruitful area for probing.

_____

_____

c. Question probe.

_____

_____

d. Statement probe.

_____

_____

C. Graduate Student, 25, to an instructor whom he trusts "I have two term papers due tomorrow. I'm giving a report in class this afternoon. My wife is down with the flu. And now I find out that a special committee wants to talk with me about my progress in the program. I've been working a couple of part-time jobs—I've got to just to keep things going. I know I'm behind, but I still think I can handle things. I hope they're not going to give me ultimatums."

a. Empathy.

_____

_____

b. Fruitful area for probing.

_____

_____

c. Question probe.

_____

_____

d. Statement probe.

_____

_____

D. A man, 49, talking to a rehabilitation counselor after an operation that has left him with one lung: "I'll never be as active as I used to be. But at least I'm beginning to see that life is still worth living. I have to take a long look at the possibilities, no matter how much they've narrowed. I can't explain it, but there's something good stirring in me."

a. Empathy.

_____

_____

b. Fruitful area for probing.

_____

_____

c. Question probe.

_____

_____

d. Statement probe.

_____

_____

E. Mark and Lisa, a married couple, both 33, after years of attempting to have children finally adopted a baby girl, Andrea. Their relationship, which up to then seemed quite good, has begun to disintegrate. Mark has made some cracks about "the stronger one in the house." Andrea has proved to be a somewhat difficult baby. Lisa feels exhausted and blames Mark for not helping her. They are both thinking about divorce now, but feel very guilty because of the child. Both of them say, "If only we had never adopted Andrea."

a. Empathy.

_____

_____

b. Fruitful area for probing.

_____

_____

c. Question probe.

_____

_____

d. Statement probe.

_____

_____

III. SUMMARIZING

At a number of points throughout the helping process it is useful for helpers to summarize or to have clients summarize the principal points of their interactions. This places clients under pressure to focus and move on. Since summarizing can place pressure on clients to "move forward," it is sometimes a form of challenge.

EXERCISE 19: SUMMARIZING AS A WAY OF PROVIDING FOCUS

This exercise assumes that trainees have been using the skills and methods of Steps I-A and I-B to help one another.

1. The total training group is divided into subgroups of three.
2. There are three roles in each subgroup: helper, client, and observer.

3. The helper spends about eight minutes counseling the client. The client should continue to explore one of the problem areas he or she has chosen to deal with in the training group.

4. At the end of four minutes, the helper summarizes the principal points of the interaction. Helpers should try to make the summary both accurate and concise. The helper can draw on past interactions if he or she is counseling a "client" whom he or she has counseled before. At the end of eight minutes or so, the helper should engage in a second summary.

5. At the end of each summary, the helper should ask the client to draw some sort of implication or conclusion from the summary. That is, the client is asked to take the next step.

6. Then both observer and client give the helper feedback as to the accuracy and the helpfulness of the summary. The summary is helpful if it moves the client toward problem clarification, goal setting, and action.

7. This process is repeated until each person in the subgroup has had an opportunity to play each role.

Example 1: It would be too cumbersome to print eight minutes of dialogue here, but consider this brief outline of a case. A young man, 22, has been talking about some developmental issues. One of his concerns is that he sees himself as relating to women poorly. One side of his face is scarred from a fire that occurred two years previous to the counseling session. He has made some previous remarks about the difficulties he has relating to women. After four minutes of interaction, the helper summarizes:

- **HELPER**: Dave, let me see if I have the main points you've been making. First, because of the scars, you think you turn women off before you even get to talk with them. The second point, and I have to make sure that this is what you are saying, is that your initial approach to women is cautious, or cynical, or maybe even subtly hostile since you've come to expect rejection somewhat automatically.
- **DAVE**: Yeah, but now that you've put it all together, I am not so sure that the hostility is all that subtle.
- **HELPER**: You also said that the women you meet are cautious with you. Some might see you as a bit nasty, I believe that was your word. Others steer clear of you because they see you as difficult to talk to and get to know. Finally, what closes the circle is that you interpret their caution as being turned off by your physical appearance.
- **DAVE**: I don't like to hear it that way, but that's what I've been saying.
- **HELPER**: If these points are fairly accurate, I wonder what implication you might see in them.
- **DAVE**: I'm the one that rejects me because of my face. Nothing's going to get better until I do something about that.

Note that the client draws an implication from the summary ("I am the primary one who rejects me") and moves on to some minimal declaration of intent ("I need to change this").

Example 2: Artemis, a 37-year-old woman, has been talking about her lack of assertive behavior in a paraprofessional helper training group. She sees clearly that his lack of assertion stands in the way of being an effective helper. She and her helper explore this theme for a few minutes and then the helper gives the following summary:

- **HELPER**. I'd like to take a moment to pull together the main points of our conversation. You're convinced that the ability to move reasonably into the life of the client is essential for you as a helper. However, this simply has not been part of your normal interpersonal style. You don't intrude into the lives of others. If anything, you are too hesitant to place demands on anyone. When you take the role of helper in training sessions, you feel awkward using even basic empathy and even more awkward using probes. As a result, you let your clients ramble and their problems remain unfocused. Outside training sessions you still see yourself as quite passive, except now you're much more aware of it. If this is more or less accurate, what implication might you draw from it?

- **TRAINEE**: When I hear it all put together like that, my immediate reaction is to say that I shouldn't try to be a counselor. But I think I would be selling myself short. No matter what career I choose to follow, I need to become more assertive. I have to learn how to take risks.

After each trainee gives his or her summary and elicits some reaction from the client, the feedback from the client and the observer should center on the accuracy and the usefulness of the summary, not on a further exploration of the client's problem. Recall that feedback is most effective when it is clear, concise, behavioral, and nonpunitive.

PART THREE

STAGE I OF THE HELPING MODEL AND ADVANCED COMMUNICATION SKILLS

The communication skills reviewed in Part 2 are critical tools; with them, you can help clients engage in the stages and steps of the helping model. But those communication skills are not the helping process itself. Part 3 focuses on Stage I of the helping model, together with its three steps. Exercises that focus on Step I-A are found in Section 7. Advanced communication skills—those related to helping clients challenge themselves—are discussed and illustrated in Chapters 9, 10, and 11. In Chapter 12, the concept of leverage—helping clients choose the right issues to work on—is explained and illustrated.

In Parts III and IV of this manual the exercises focus on the stages and steps of the helping process together with the advanced communication skills needed to help clients get value from each session and make progress. In Part III the focus is on the three steps of Stage I and steps of the helping model. These exercises will enable you help your clients:

- identify and clarify problem situations and missed opportunities.
- challenge themselves to develop new and more useful perspectives on themselves and their problems.
- work on key issues, that is, issues that will make a difference in their lives.
- commit themselves to work on these issues in pragmatic, action-oriented ways.

As indicated in the text, the steps or tasks of Stage I are not only interrelated but apply to all stages and steps of the helping process. That is, throughout the counseling process clients need to be helped tell their stories, challenge themselves, and work on things that make a difference.

Section 7
STEP I-A: HELPING CLIENTS TELL THEIR STORIES

The communication skills reviewed in Part Two are tools you need to deliver the stages and steps of the helping model. Stage I deals with problem identification and clarification. Step A in this stage deals with helping clients tell their stories, that is, discuss the problem situations, anxieties, and concerns that bring them (or get them sent) to the helper in the first place. Read Chapter Seven in *The Skilled Helper* before doing the exercise in this section.

EXERCISE 20: PREPARING TO TELL YOUR STORY—PROBLEMS

Since discussing the problems and unused opportunities that could affect the effectiveness of your helping is an important part of the training, this and the following sentence-completion exercise will help you to choose issues that you would like and be willing to discuss.

Do these sentence-completion exercises quickly. They may help you expand in more specific ways what you have learned about yourself in the preceding exercises.

1. My biggest problem is _____

_____

2. I'm quite concerned about _____

_____

3. One of my other problems is _____

_____

4. Something I do that gives me trouble is _____

_____

5. Something I fail to do that gets me into trouble is _____

_____

6. The social setting of life I find most troublesome is _____

_____

7. The most frequent negative feelings in my life are _____

_____

8. These negative feelings take place when _____

_____

9. The person I have most trouble with is _____

_____

10. What I find most troublesome in this relationship is _____

_____

11. Life would be better if _____

_____

12. I tend to do myself in when I _____

_____

13. I don't cope very well with _____

_____

14. What sets me most on edge is _____

_____

15. I get anxious when _____

_____

16. A value I fail to put into practice is _____

_____

17. I'm afraid to _____

_____

18. I wish I _____

_____

19. I wish I didn't _____

_____

20. What others dislike most about me is _____

_____

21. What I don't seem to handle well is _____

_____

22. I don't seem to have the skills I need in order to _____

_____

23. A problem that keeps coming back is _____

24. If I could change just one thing in myself it would be _____

_____

EXERCISE 21: PREPARING TO TELL YOUR STORY—UNUSED OPPORTUNITIES

This sentence-completion exercise mirrors the preceding one but deals with unused or underused resources and opportunities rather than problems. Undeveloped resources and opportunities can affect the quality of your helping.

1. One thing I like about myself is _____

_____

2. One thing others like about me is _____

_____

3. One thing I do very well is _____

_____

4. A recent problem I've handled very well is _____

_____

5. When I'm at my best I _____

_____

6. I'm glad that I am able to _____

_____

7. Those who know me know that I can _____

_____

8. A compliment that has been paid to me recently is _____

_____

9. A value that I try hard to practice is _____

_____

10. An example of my caring about others is _____

_____

11. People can count on me to _____

_____

12. They said I did a good job when I _____

_____

13. Something I'm handling better this year than last is _____

_____

14. One thing that I've overcome is _____

_____

15. A good example of my ability to manage my life is _____

_____

16. I'm best with people when _____

_____

17. One goal I'm presently working toward is _____

_____

18. A recent temptation that I managed to overcome was _____

_____

19. I pleasantly surprised myself when I _____

_____

20. I think that I have the guts to _____

_____

21. If I had to say one good thing about myself I'd say that I _____

_____

22. One way I successfully control my emotions is _____

23. One way in which I am very dependable is _____

24. One important thing I intend to do within the next two months is _____

EXERCISE 22: HELPING ONE ANOTHER TELL YOUR STORIES

In the exercises you have done to this point you have already been helping clients tell their stories through the use of empathy and probes. It is time now to switch from writing to dialogue.

1. Form groups of three.
2. In each group there are three roles: client, helper, and observer.
3. Before the dialogue between client and helper begins, the trainee in the client role gives a brief summary of the problem situation or unused opportunity he or she is going to explore.

Example: Enrico, the trainee playing the role of client, gives the following summary:

> I'm going to discuss a problem I'm having with my parents. I'm an only child and they simply won't let go. Don't get me wrong. They're very nice about it. But I'm 20. We're not a wealthy family. So, for the time being, I have to live at home. Also I believe that becoming my own person is directly related to becoming an effective helper.

4. The helper begins by saying: "Against that background, Enrico, what would you like to focus on today."
5. Then client and helper engage in a helping dialogue. The helper, using whatever combination of empathy and probes he or she sees necessary, helps the client tell his or her story clearly and in useful detail.
6. After about six minutes the observer, in timekeeper mode, stops the dialogue.
7. The client gives the helper feedback on how effectively he or she has aided the story telling process. (Since giving feedback to your fellow learners is an important part of the training program, see the section below on hints for giving effective feedback.)
8. Then the observer complements what the client says with his or her feedback.
9. The process continues until each member has had the opportunity to play each role.
10. Then proceed to a second round. In the second round the trainee in the client role is asked to summarize the first dialogue and then move on from there.

GIVING FEEDBACK TO SELF AND OTHERS

If you are working with a learning partner or if you are in an experiential training group, you will be called on to give feedback both to yourself and to your co-learners on how well you are learning and using helping methods and skills. Giving feedback well is an art. Here are some guidelines to help you develop that art.

1. **Keep the goal of feedback in mind.** In giving feedback, always keep the generic goal of feedback in mind which is to help the other person (or yourself, in the case of self-feedback) do a better job. Improved performance is the desired outcome. Applied to helping, this means providing the kind of feedback to yourself and your co-learners that will help you become better helpers. Feedback will help you learn every stage and step of the model.

2. **Give positive feedback.** Tell your co-learners what they are doing well. This reinforces useful helping behaviors. "You leaned toward him and kept good eye contact even though he became very intense and emotional. Your behavior sent the right I'm-still-with-you message."

3. **Don't avoid corrective feedback.** To learn from our mistakes we must know what they are. Corrective feedback given in a caring way is a powerful tool for learning. "You seem reluctant to challenge your clients. For instance, Sam [the client] didn't fulfill his contract from the last meeting and you let it go. You fidgeted when he said he didn't get around to it."

4. **Be specific.** General statements like, "I liked your style in challenging your client," or "You could have been more understanding," are not helpful. Change them to specific descriptions such as, "Your challenge was helpful because you pointed out how self-defeating her internal conversations with herself are and you hinted at ways she could change these conversations." Or, "Your tone was harsh and you did not give him a chance to reply to what you were saying."

5. **Focus on behavior rather than traits.** Point out what the helper does or fails to do. Do not focus on traits or use labels such as, "You showed yourself to be a leader." or "You're still a bumbler." Avoid using negative traits such as "lazy," "a slow learner," "incompetent," "manipulative," and so forth. This is just name calling and creates a negative learning climate in the group. The following statements deal with specific behaviors rather than traits. "You let him criticize you without becoming defensive; he listened to you better after that." or "You did not catch her core message; in fact, you seem to have difficulty in listening well enough to catch your clients' core messages." Such statements deal with specific behaviors rather than traits.

6. **Indicate the impact of the behavior on the client.** Feedback should help counselors-to-be interact more productively with clients. It helps, then, to indicate the impact of the helper's behavior on the client. "You interrupted the client three times in the space of about two minutes. After the third time, she spoke less intensely and switched to safer topics. She seemed to wander around."

7. **Provide hints for better performance.** Often helpers, once they receive corrective feedback, know how to change their behavior. After all, they are learning how to help in the training program. Sometimes, however, if the helper agrees with the feedback but does not know how to change his or her behavior, suggestions or hints on how to improve performance are useful. These, too, should be specific and clear. "You are having trouble providing your clients with empathy because you allow them to talk too long. When they go on and on, they make so many points that you don't know which to respond to. Try 'interrupting' your clients gently so that you can respond to key messages as they come up."

8. **Be brief.** Feedback that is both specific and brief is most helpful. Long-winded feedback proves to be a waste of time. A helper might need feedback on a number of points. In this case, provide feedback on one or two points. Give further feedback later on. In general, do not overload your co-learners with too much feedback at one time.

9. **Use dialogue.** Feedback is more effective if it takes place through a dialogue between the giver and receiver, a brief dialogue, of course. This gives the receiver an opportunity to clarify what the feedback giver means and to ask for suggestions if he or she needs them. A dialogue helps the receiver better "own" the feedback.

EXERCISE 23: LINKING STORYTELLING TO ACTION

As noted in the text, clients need to engage in "little" actions throughout the helping process that get them moving in the right direction even before formal goals and action strategies are established. These little actions are signs of clients' commitment to the process of constructive change. In this exercise you are asked to put yourself in the place of the client and come up with some posible actions that the client might take to move forward in some way that will contribute to managing the problem situation or developing the unused opportunity better. It is not that you are going to tell the client what to do. But helping clients adopt an action orientation to their problem situations is central to helping.

1. Listen to the story.
2. If you were the client, what are two common-sense actions you might consider taking at this stage in order to move forward. Even if "big" actions seem premature, what "little" actions might you take?
3. Indicate the reason for each possible action.

Example

A seventh-grade boy talking to a teacher he trusts (all this is said in a halting voice and he does not look at the teacher):

> Something happened yesterday that's bothering me a lot. I was looking out the window after school. It was late. I saw two of the guys, the bullies, beating up on one of my best friends. I was afraid to go down there. . . . A coward. . . . I didn't tell anyone, I didn't do anything.

If you were the client: Two possible client actions and reason for each: talking the incident through with his friend and perhaps apologizing (reason: to reestablish relationship with his friend and to show basic decency); figuring out how he wants to handle similar situations in the future (reason: to learn something from this painful experience).

Note that this exercise is meant to help you develop an action orientaton throughout the helping process. It is not suggested that you should be telling clients what to do. Rather, through probing and challenging, you can help them identify actions they might take even early on.

1. A young woman talking to a counselor in a center for battered women. She has been seen by other counselors on two different occasions: "This is the third time he's beaten me up. I didn't come before because I still can't believe it! We're married only a year. After we got married, he began ordering me around in ways he never did before we got married. He'd get furious if I questioned him. Then he began shoving me if I didn't do what he wanted fast or right. And I just let him do it! I just let him do it! (She breaks down and sobs.) And now three beatings in less than two months. Oh God, what's happened?"

Two possible client actions and a reason for each.

65

2. A girl, 15, talking to a psychologist at a time when her parents are involved in a divorce case: "I still want to do something to help, but I can't. I just can't! They won't let me. When they would fight and get real mean and were screaming at each other, I'd run and try to get in between them. One or the other would push me away. They wouldn't pay any attention to me at all. They're still pushing me away. They don't care how I think or feel or what happens to me! My mother tells me that kids should stay out of things like this."

Two possible client actions and a reason for each.

_____

_____

_____

3. A man, 25, is talking with a fellow trainee outside the group: "In yesterday's group session I was watching Leonard [the trainer] give Peggy feedback. He does it very well. But I was also saying to myself, 'Why isn't he that helpful and that careful with me?' I want the same kind of feedback, but he doesn't say much to me at all. I'm as active as anyone else in the group. I volunteer for the demonstrations in front of the group. Well, you know that. I don't know why you he passes me by. Or is all of this just my imagination?"

Two possible client actions and a reason for each.

_____

_____

_____

4. A woman, 63, in a hospital dying of cancer, is talking to a member of the pastoral counseling staff: "I can understand it from my children, but not from my husband. I know I'm dying. But he comes here with a brave smile every day, hiding what he feels. We never talk about my dying. I know he's trying to protect me, but it's so unreal. I don't tell him that his constant cheerfulness and his refusal to talk about my sickness are actually painful to me. (She shakes her head.) I'm being careful of _him_!"

Two possible client actions and a reason for each.

_____

_____

_____

5. A man, 53, is talking to a counselor a few months after the sudden death of his wife. His two children are married and living in distant towns. "I miss her so. The house seems so empty. At my job I work alone on computer programming. There's no one I talk to at work. Now there's no one at home. I walk around

the house thinking of how I was with her in each room. At night sometimes I sit in the dark thinking of nothing. We had few friends, so no one calls. And I haven't seen either of the kids since the funeral."

Two possible client actions and a reason for each.

_____

_____

_____

The following are supplemental exercises.
They should be filled in only if the you have not yet gotten the point or
feel that further written practice will help you master this part of the helping model.

A. A woman, 35, with two children, one four, one six, whose husband has deserted them, is talking to a social worker: "He's not sending me any money. I don't even know where he is. They're asking me for the rent and telling me that I have to find some way of coming up with it. I've been to two different agencies and filled out all sorts of forms, but I don't have any money or food stamps yet. I've been getting food from my mother, but she's really got next to nothing. What am I supposed to do? I'll work, but who's to take care of the kids. I asked all around and there's no day-care center anywhere near here."

Two possible client actions and a reason for each.

_____

_____

_____

B. A freshman in college talking to a counselor toward the end of his first year: "One week I find myself studying hard, working on the school paper, going to a talk on foreign affairs. The next week I'm boozing it up, looking around for a hot sex partner, and playing cards all day with the boys. It's like being two different people who don't even know each other! I like being in college, away from home and all that. It's such a mixed bag. For the most part, I like college, but all the stuff I like the best is outside the classroom. I don't know what I want."

Two possible client actions and a reason for each.

_____

_____

_____

C. A man, 70, arrested for stealing funds from the company where he has worked for 25 years, talking to an assigned social worker. This is their second session: "To tell you the truth, it's probably a good thing I've been caught. I've been stealing on and off for the last five or six years. It's been a game. It soaked up my energies, my attention, distracted me from thinking about getting old. Now I'm saying to myself: 'You old fool, what're you running from?' I've been forcing myself to try to make sense out of my life. You're probably thinking: 'It's about time, old guy.' I'm thinking it's as good a time as any."

Two possible client actions and a reason for each.

_____

_____

_____

D. A woman, 37, married, with an unwanted pregnancy; she has two children, one in the seventh and one in the eighth grade; she is talking to another woman, her closest friend who is a good common-sense counselor: "Ellen, I just don't know what to do. I've talked to my pastor, but I knew what he was going to say. He wasn't much help at all. Oh, God, I don't want another child! Not now! A couple of people I know just assume I'll have an abortion. That's what they'd do. I won't have an abortion, I just won't. But I don't want to have to restructure my life. I've had the children I wanted!"

Two possible client actions and a reason for each.

_____

_____

_____

E. A minister who has been having an affair with one of his parishioners, talking to another minister: "I've never known anyone like her before. It was as if it didn't make any difference that both of us were married. I've never experienced such strong emotion. I can tell myself exactly what I should do. But I don't do it. We avoid talking about where all this is going to lead. I know in the back of my mind that my family and career and all that are on the line, but I keep it pushed back. There's doom on the horizon, but the present is so damn full!"

Two possible client actions and a reason for each.

_____

_____

_____

Section 8
RELUCTANT AND RESISTANT CLIENTS

In order to recognize reluctance and resistance in clients, it helps to first be able to recognize them within yourself and learn how to manage them in your own life. Read Chapter Eight before doing the following exercise.

EXERCISE 24: MY OWN RELUCTANCE

Since growth and living more effectively requires work, all of us express reluctance from time to time. This exercise targets your own reluctance to become more than you are. In Sections 1, 6, and 7 you have been asked to review some of your own problems, unused opportunities, and developmental tasks. These exercises provide fodder for this exercise.

1. Review the problems, unused or underused opportunities and resources, and developmental challenges in your life.
2. Identify three areas or ways in which you have been reluctant to grow or change. Spell them out in some detail.

Example. Michela, 19, a first-year student in clinical psychology, has this to say about developing a sense of responsibility:

> The more I learn bout the world, the more I see how comparatively easy life has been for me. Everything has been given to me. I'm a product of the American dream. I had an easy time in high school, but the first months of graduate school have been jarring. No one is going to give me anything. As I look down the road I see that this experience more accurately reflects the real world. Without getting into useless guilt trips, I realize I'm spoiled. I also realize that old habits do not die easily. I still expect things to be handed to me on a platter. I'm reluctant to face up to the realities of imposed schedules, competition, demanding work, and budgets. It is very difficult for me to accept that some of my learning is gong to come from making mistakes, being challenged, and doing it again.

Area #1 of reluctance.

_____

_____

_____

_____

_____

Area #2 of reluctance.

_____

_____

_____

_____

_____

_____

Area #3 of reluctance.

_____

_____

_____

_____

_____

_____

3. Choose one of the areas and indicate what you might do in order to overcome your reluctance.

_____

_____

_____

_____

_____

4. As you share these instances of personal reluctance with a learning partner or the entire training group, see if you can identify any underlying themes that run through all three areas.

EXERCISE 25: MY OWN EXPERIENCE OF RESISTANCE

Recall that resistance is a reaction to someone's trying to *force* us to change when we don't want to. We may even think that the other person has a point, but the fact that he or she is telling us what to do causes us to react rather than respond.

1. Recall two instances when you resisted growth because you felt you were being forced into it.
2. Indicate how that negative experience could have become a positive one for you.

Example. Manfred, a 24-year-old graduate student in social work, recalls being badgered by a former girl friend:

> Elisa was out to reform me. It's not that I was not in need of reform during my college years. But the way she went about it turned me off. For instance, I drank too much. But she lectured me private. She embarrassed me in front of my friends—her way of getting my attention. Before gong out to major events like the homecoming dance, she laid down rules and made me promise to keep them. Otherwise she wouldn't go out with me. Things like that.

When asked how that negative experience could have been turned into a positive one, he had this to say among other things:

> The substance of what she was saying actually made sense. To put it in one phrase, it would have gone better if she had *backed off* a bit, quite a bit at times. For instance, instead of laying down rules to prevent disasters, she could have quietly left the scene if they did occur. If she had said that she was no longer having fun and was going home and then just did it a couple of times, I think that would have been a real walk up call for me.

Now do the same.

Instance a.

_____

_____

_____

_____

_____

Indicate how that negative experience could have become a positive one for you.

_____

_____

_____

_____

_____

Instance b.

_____

_____

_____

_____

_____

Indicate how that negative experience could have become a positive one for you.

_____

_____

_____

_____

_____

Section 9
STEP I-B: I. THE NATURE OF CHALLENGING—
HELPING CLIENTS CHALLENGE THEMSELVES

Chapter Nine in *The Skilled Helper* discusses the role of challenging in the helping process. Helping clients challenge themselves both to participate actively in the helping process and to act on what they learn adds great value. While challenge without understanding is abrasive, understanding without challenge is anemic. Before doing the following exercise, read Chapter Nine.

We begin with an exercise in self-challenge. Many of the clients you see will be struggling with developmental concerns like those you have struggled with or are currently working on. Like you, they will have both strengths and "soft spots" in coping with the developmental tasks of life.

EXERCISE 26: REVIEWING SOME BASIC DEVELOPMENTAL TASKS

In this exercise you are asked to consider your experience with ten major developmental tasks of life. First reflect on your experience in these developmental areas and then apply what you have learned to your role as a helper of others. Use extra paper as needed.

1. **COMPETENCE. What do I do well?** Do I see myself as a person who is capable of getting things done? Do I have the resources needed to accomplish goals I set for myself? In what areas of life do I excel? In what areas of life would I want to be more competent than I am?

Strengths   **Soft Spots**

_____   _____

_____   _____

_____   _____

_____   _____

2. **AUTONOMY: How well do I make it on my own**? Can I get things done on my own? Do I avoid being overly dependent or independent? Am I reasonably interdependent in my work and social life? When I need help, do I find it easy to ask for it? In what social settings do I find myself most dependent? counterdependent? independent? interdependent?

Strengths   **Soft Spots**

_____   _____

_____   _____

_____   _____

_____   _____

3. **VALUES: What do I believe in and prize?** What are my principal values? Do I allow for reasonable changes in my value system? Do I put my values into practice? Do any of the values I hold conflict with others? In what social settings do I pursue the values that are most important to me?

Strengths   **Soft Spots**

_____   _____

_____   _____

_____   _____

_____   _____

4. **IDENTITY: Who am I in this world?** Do I have a good sense of who I am and the direction I'm going in life? Do the ways that others see me fit with the ways in which I see myself? Do I have some kind of center that gives meaning to my life? In what social settings do I have my best feelings for who I am? In what social settings do I lose my identity? In what ways am I confused or dissatisfied with who I am?

Strengths   **Soft Spots**

_____   _____

_____   _____

_____   _____

_____   _____

5. INTIMACY. What are my closer relationships like? What kinds of closeness do I have with others? To what extent are there degrees of closeness in my life—acquaintances, friends, and intimates? What is my life in my peer group like? How well do I get along with others? What concerns do I have about my interpersonal life?

Strengths   **Soft Spots**

_____   _____

_____   _____

_____   _____

_____   _____

6. SEXUALITY. Who am I as a sexual person? To what degree am I satisfied with my sexual identity, my sexual preferences, and my sexual behavior? How do I handle my sexual needs and wants? What social settings influence the ways I act sexually?

Strengths   **Soft Spots**

_____   _____

_____   _____

_____   _____

_____   _____

7. LOVE, MARRIAGE, FAMILY. What are my deeper commitments like? What is my marriage like? How do I relate to family and relatives? How do I feel about the quality of my family life? If not married, in what ways do I look forward to marriage? What misgivings do I have?

Strengths

Soft Spots

_____

_____

_____

_____

8. CAREER. What is the place of work in my life? How do I feel about the way I am preparing myself for a career or the career I am currently pursuing? What do I get out of work? What am I like in the workplace? How does it affect me? What impact do I have there?

Strengths

Soft Spots

_____

_____

_____

_____

9. INVESTMENT IN THE WIDER COMMUNITY. How big is my world? How do I invest myself in the world outside of friends, work, and the family? What is my neighborhood like? Do I have community, civic, political, social involvements or concerns? In what ways am I optimistic about the world? In what ways am I cynical?

Strengths

Soft Spots

_____

_____

_____

_____

10. LEISURE. What do I do with my free time? Do I feel that I have sufficient free time? How do I use my leisure? What do I get out of it? In what social settings do I spend my free time?

Strengths	**Soft Spots**
_____	_____
_____	_____
_____	_____
_____	_____

Many of the clients you see will be struggling with similar developmental concerns. Like you, they will have both strengths and "soft spots" in these areas.

1. Pick two of the ten developmental areas. Try to pick areas that will make a difference in your work as a counselor.
2. Indicate one or two new perspectives generated because of this exercise.
3. Outline the kinds of actions these new perspectives can drive.

Example. Mickey, a 20-year-old undergraduate majoring in psychology, had this to say:

- Developmental areas: "I chose identity and investment in the wider community.
- New perspective: "There is a huge difference between identity and what I now see as individualism. I think I've been on the selfish, me-first, I'm-what's-important track. I think little about the wider community and belonging to it."
- Action implications: "First of all, I have to clear up in my mind the difference between knowing who I am, what I stand for, and where I'm going and my me-first attitudes. Second, I'd like to begin to take a critical look at the messages the society of which I am a part is screaming at me through music, movies, television, peer interaction—the whole package. A lot of the messages would keep me an adolescent a an adult and I don't want to remain an adolescent."

Now explore two areas for yourself.

a. First developmental area.

_____

New perspectives.

_____

_____

_____

Link these new perspectives to action.

_____

_____

_____

_____

b. Second developmental area.

_____

New perspectives.

_____

_____

_____

Link these new perspectives to action.

_____

_____

_____

_____

4. Share your learnings with a partner from your training group.

EXERCISE 27: IDENTIFYING DIFFERENT AREAS NEEDING CHALLENGE

Challenge focuses on dysfunctional mindsets and related actions, internal behavior, and external behavior. There are various ways in which clients need to be helped challenge themselves in order to develop the kinds of new perspectives that lead to constructive problem-managing and opportunity-developing action. In this exercise you are asked to do the following:

1. Read the case and determine in what way or in what area this client might benefit from challenge.
2. Sketch out briefly what some helpful new perspectives might look like.
3. Share your observations with a learning partner. Note the similarities and the differences among your suggestions.
4. Discuss how you might proceed in each case, finding ways to help these clients challenge themselves.

Example: Hannah, a new counselor working in a hospital setting for a number of months, has begun to confide in one of her colleagues who has worked in health-care facilities for about seven years. Hannah has intimated, however tentatively, that she feels that some of the medical doctors tend to dismiss her because she is new, because she is a counselor and not a medical practitioner, and because she is a woman. One doctor has been particularly nasty, even though he masks his abusiveness under his brand of "humor." One day her colleague says to her, "Tell 'em what you think!" Hannah replies: "That will only make them worse. I think it more important to stand up for my profession by just being who I am and helping patients as much as possible. Their meanness is a punishment in itself. Anyway, it's a game that doctors play with everyone else in the hospital. That kind of thing has been going on forever. It's not such a big deal anyway."

- **What mindsets and behavior might be challenged here?** Under the guise of "reasonableness," Hannah seems to be into excuse making. She might be helped to explore her reasons for not doing anything about the situation. There could be reasons under her reasons that might reveal other issues in the problem situation.
- **What would some helpful new perspectives look like?** Hannah might be helped to see that accepting the status quo goes counter to the kind of professionalism that should characterize the hospital and that she prides herself in. She may also be helped to see that the hospital's caste system might interfere with team approaches to helping patients.
- **What actions might such new perspectives lead to?** Hannah might bring the core issues before the hospital senate, she might quietly lobby some of her colleagues to see what kind of consensus there is, she might talk to the abusive doctor privately and let him know how she feels, she might confide in one of the doctors who seems to be free of these prejudices in order to see what course of action she might follow, and so forth.

It goes without saying that the best way of doing this exercise is through actual interaction with these clients. At this stage, however, you are being asked to do some educated guessing. Continue with the following cases.

1. Dan is in a coaching/counseling session with his supervisor, June. A project is behind schedule and a key customer is complaining. Dan was chosen to be the project leader because he had manifested "leadership" qualities in his work. He says something like this: "Well, this is one of the first team-focused projects we've had. We're not used to working like this. The key people in design and engineering have been slow to come to the table. They keep telling me that they can't let other projects slip. I know the customer wants the prototype by the end of the month, but that's his schedule, not necessarily reality. He's so pushy. Lately, it's hard even taking his calls."

a. What mindsets and related behavior might be challenged here?

_____

_____

_____

b. What would some helpful new perspectives look like?

_____

_____

_____

c. What actions might such a perspective lead to?

_____

_____

_____

2. Claudette and Jim have tried unsuccessfully to have children. Doctors have examined both of them and now offer little hope even with new medical techniques. They have both been seeing a counselor, but in this session Claudette, who is much less reconciled to their fate than Jim is, is seeing the counselor alone. She says, "It's just not fair. Doctors talk about infertility as a clinical phenomenon, not a human reality. Jim and I love one another. We want to have children. We want the sex we have together to lead to life. We'd make very good parents. We've done everything the doctors have told us to do, and still nothing."

a. What mindset or behavior might be challenged here?

_____

_____

_____

b. What would a helpful new perspective look like?

_____

_____

_____

c. What actions might such a perspective lead to?

_____

_____

_____

3. Len, married with three teenage children, lost all the family's savings in a gambling spree. Under a lot of pressure from both his family and his boss, he started attending Gamblers Anonymous meetings. He seemed to recover, that is, he stopped betting on horses and ball games. A couple of years went by and Len stopped going to the GA meetings because he "no longer needed them." He even got a better job and was recovering financially quite well. But his wife began to notice that he was on the phone a great deal with his broker. She confronted him and he reluctantly agreed to a session with one of the friends he made at GA. He says, "She's worried that I'm gambling again. You know, it's really just the opposite. When I was gambling I was financially irresponsible. I lost our future on horses and ball games. But now

I'm taking a very active part in creating our financial future. Every financial planner will tell you that investing in the market is central to sound financial planning. I'm a doer. I'm taking a very active role."

a. What mindsets and behavior might be challenged here?

_____

_____

_____

b. What would a helpful new perspective look like?

_____

_____

_____

c. What actions might such a perspective lead to?

_____

_____

_____

4. Stepan, a refugee who had been brutalized by a political regime in his native country, has been seeing a counselor in a center that specializes in helping victims of torture. He has been in the country for two years. He has a decent job. Even though he is now secure, at least in some sense of that term, he has found it very difficult to establish relationships with other refugees, with members of the immigrant community, or with native-born Americans. He intimates that he is quite lonely, but dismisses loneliness as "inconsequential" when the counselor brings it up. In one session, he says: "Let's not talk about my so-called loneliness when there is a great world loneliness out there. The loneliness that brutality creates, now that's something that counts. I fill my days quite well. When it is time, I'll seek out others, but I need people who can see the world as I see it. As it really is."

a. What mindsets and behavior might be challenged here?

_____

_____

_____

b. What would a helpful new perspective look like?

_____

c. What actions might such a perspective lead to?

_____

_____

_____

Section 10
STEP I-B: II. SPECIFIC CHALLENGING SKILLS

Specific challenging skills, as outlined in Chapter Ten of *The Skilled Helper*, include advanced empathy, information sharing, helper self-disclosure, immediacy, suggestions, advice, and directives, confrontation. The purpose of these skills is to help clients get in touch with blind spots and develop the kind of new perspectives or behavioral insights needed to complete the clarification of a problem situation and to move on to developing new scenarios, setting problem-managing goals, developing strategies, and moving to action. That is, challenging skills can be used throughout and serve the entire helping process. For reasons outlined in the text, exercises are limited to advanced empathy, information sharing, helper self-disclosure, and immediacy. Read the relevant section in the text before doing each set of exercises.

ADVANCED EMPATHY: THE MESSAGE BEHIND THE MESSAGE

Advanced empathy, described most simply, means sharing educated **hunches** about clients and their overt and covert experiences, behaviors, and feelings that you feel will help them see their problems and concerns more clearly and help them move on to developing new scenarios, setting goals, and acting. Such hunches, of course, must be based on the helper's interactions with the client in which active listening, understanding, empathy, and probing play a large part. The following section is borrowed from the text.

- Hunches can help clients see the **bigger picture**. In this example, the counselor is talking to a client who is at odds with his wife's brother: "The problem doesn't seem to be just your attitude toward your brother-in-law any more; your resentment seems to be spreading to his friends. How do you see it?"
- Hunches can help clients see what they are expressing **indirectly** or merely **implying**. Example: The counselor is talking to a client who feels that a friend has let her down: "I think I might also be hearing you say that you are more than disappointed—perhaps a bit hurt and angry."
- Hunches can help clients draw **logical conclusions** from what they are saying. A manager is counseling one of his team members: "From all that you've said about her, it seems that you are also saying that right now you resent having to work with her at all. I know you haven't said that directly. But I'm wondering if you are feeling that way about her."
- Hunches can help clients open up areas they are only **hinting** at. In this case a school counselor is talking to a senior in high school: "You've brought up sexual matters a number of times, but you

81

haven't pursued them. My guess is that sex is a pretty important area for you but perhaps pretty touchy, too."

- Hunches can help clients see things they may be **overlooking**. A counselor is talking to a minister: "I wonder if it's possible that some people take your wit too personally, that they see it, perhaps, as sarcasm rather than humor."
- Hunches can help clients identify **themes**. In this case a counselor is talking to a woman who has been abused by her husband: "If I'm not mistaken, you've mentioned in two or three different ways that it is sometimes difficult for you to stick up for your own legitimate rights. For instance . . . ."
- Hunches can help clients **own** more fully partially-owned experiences, behaviors, and/or feelings. Example: "You sound as if you have already decided to quit. Or is that overstating the case?"

Hunches should be based on your experience of your clients—their experiences, behaviors, and emotions both within the helping sessions themselves and in their day-to-day lives. Do not base your hunches on "deep" psychological theories. Later on, you will be asked to identify the experiential and behavioral clues on which your hunches are based.

EXERCISE 28: ADVANCED ACCURATE EMPATHY—HUNCHES ABOUT ONESELF

One way to get an experiential feeling for advanced empathy is to explore *at two levels* some situation or issue in your own life that you would like to understand more clearly. One level of understanding could be called the surface level. The second could be called a deeper level. Often reality is found at the deeper level.

1. Read the examples given below.

Example 1: A man, 25, in a counselor training program is having his ups and downs in the training group. He is getting second thoughts about his ability and willingness to get close to others.

- **Level-1 understanding:** "I like people and I show this by my willingness to work hard with them. For instance, in this group I see myself as a hard worker. I listen to others carefully and I try to respond carefully. I see myself as a very active member of this group. I take the initiative in contacting others. I like working with the people here."
- **Below-the-surface understanding:** "If I look closer at what I'm doing here, I realize that underneath my hardworking and competent exterior, I am uncomfortable. I come to these sessions with more misgivings than I have admitted, even to myself. My hunch is that I have some fears about human closeness. I am afraid, both here and in a couple of relationships outside the group, that someone is going to ask me for more than I want to give. This keeps me on edge here. It keeps me on edge in a couple of relationships outside."

Example 2: A woman, 33, in a counselor training group, sees that her experience in the group is making her explore her attitude toward herself. It might not be as positive as she thought. She sees this as something that could interfere with her effectiveness as a counselor.

- **Level-1 understanding:** "I like myself. I base this on the fact that I seem to relate freely to others. There are a number of things I like specifically about myself. I'm fairly bright. And I think I can use my intelligence to work with others as a helper. I work hard. I'm demanding of myself, but I don't place unreasonable demands on others."

- **Below-the-surface understanding**: "If I look more closely at myself, I see that when I work hard it is because I feel I have to. My hunch is that I-have-to' counts more in my hard work than I-want-to. I get pleasure out of working hard, but it also keeps me from feeling guilty. If I don't work hard enough as I define the term, then I can feel guilty or down on myself. I am beginning to feel that there is too much of the I-must-be-a-perfect-person in me. I judge myself and others more harshly than I care to think."

2. Choose some issue, topic, situation, or relationship that you have been investigating and which you would like to understand more fully with a view to taking some kind of action on it. As usual, choose issues that you are willing to share with the members of your training group and try to choose issues that might affect the quality of your counseling.
3. Briefly describe the issue, as in the examples.
4. Then give a "surface-level" description of the issue.
5. Next, share some hunch you have about yourself that relates to that issue, a hunch that "goes below the surface," as it were. This will help you get in touch with possible blind spots. Try to develop a new perspective on yourself and that issue, one that might help you see the issue more clearly so that you might begin to think of how you might act on it.

Area # 1.

_____

_____

_____

Level-1 understanding.

_____

_____

_____

Below-the surface understanding.

_____

_____

_____

_____

Area # 2.

_____

_____

_____

Level-1 understanding.

_____

_____

_____

Below-the surface understanding.

_____

_____

_____

_____

_____

6. Share you insights with your learning partner.

EXERCISE 29: THE DISTINCTION BETWEEN BASIC AND ADVANCED EMPATHY

In this exercise, assume that the helper and the client have established a good working relationship, that the client's concerns have been explored from his or her perspective, and that the client needs to be challenged to see the problem situation from some new perspective or to act on the insights that have been developed.

1. In each instance, imagine the client speaking directly to you.
2. First respond to what the client has just said with basic empathy.
3. Next, formulate one or two hunches about this person's experiences, behaviors, or feelings, hunches that, when shared, would help promote the kind of new perspectives that drive problem-managing action. Use the material in the "context" section together with the client's statement to formulate your hunches. Ask yourself: "On what clues am I basing this hunch?"
4. Finally, respond with some form of advanced empathy, that is, share some hunch that you believe will be useful for him or her.

Example: A man, 48, husband and father, is exploring the poor relationships he has with his wife and children. In general, he feels that he is the victim, that his family is not treating him right (that is, like many clients, he emphasizes his experience rather than his behavior). He has not yet examined the implications of the ways he behaves toward his family. At this point he is talking about his sense of humor. He says:

> For instance, I get a lot of encouragement for being witty at parties. Almost everyone laughs. I think I provide a lot of entertainment, and others like it. It also works on my job. But this is another way I seem to flop at home. When I try to be funny, my wife and kids don't laugh, at least not much. At times they even take my humor wrong and get angry. I actually have to watch my step in my own home.

a. **Basic empathy.** "It's irritating when your own family doesn't seem to appreciate what you see as one of your real talents, one that others do appreciate."
b. **Hunch.** The family wants a husband and father, not a humorist. His humor, especially at home, is not as harmless as he thinks.
c. **Advanced empathy.** "I wonder whether their reaction to you could be interpreted differently. For instance, they might not want an entertainer at home, but just a husband and father. You know, just you."

1. A first-year engineering graduate student has been exploring his disappointment with himself and with his performance in school. His father is a successful engineer, but did not pressure his son into engineering. He has explored such issues as his dislike for the school and for some of the teachers with his counselor. He says: "I just don't have much enthusiasm. My grades are just okay, maybe even a little below par. I know I could do better if I wanted to. I don't know why my disappointment with the school and some of the faculty members can get to me so much. It's not like me. Ever since I can remember—even in primary school, when I didn't have any idea what an engineer was—I've wanted to be an engineer. Theoretically, I should be as happy as a lark because I'm in a graduate school with a good reputation, but I'm not."

a. Basic empathy.

_____

_____

b. Your hunch and your reason for it.

_____

_____

_____

c. Advanced empathy.

_____

2. This man, now 64, retired early from work when he was 62. He and his wife wanted to take full advantage of the "golden" years. But his wife died a year after he retired. At the urging of friends he has finally come to a counselor. He has been exploring some of the problems his retirement has created for him. His two married sons live with their families in other cities. In the counseling sessions he has been dealing somewhat repetitiously with the theme of loss. He says: "I seldom see the kids. I enjoy them and their families a lot when they do come. I get along real well with their wives. But now that my wife is gone . . . (pause) . . . and since I've stopped working . . . (pause) . . . I seem to just ramble around the house aimlessly, which is not like me at all. I suppose I should get rid of the house, but it's filled with a lot of memories—bittersweet memories now. There were a lot of good years here."

a. Basic empathy.

_____

_____

b. Your hunch and your reason for it.

_____

_____

_____

c. Advanced empathy.

_____

_____

_____

3. A single woman, 33, is talking to a psychiatrist about the quality of her social life. She has a very close friend, Ruth, on whom she has become somewhat dependent. She is exploring the ups and downs of this relationship. This is the third session. During the sessions she comes on a bit loud and somewhat aggressive. She says: "Ruth and I are on again off again with each other lately. When we're on, it's great. We have lunch together, go shopping, all that kind of stuff. But sometimes she seems to click off. You know, she tries to avoid me. But that's not easy to do. (She laughs.) I keep after her. She's been pretty elusive for about two weeks now. I don't know why she runs away like this. I know we have our differences. But our differences don't ordinarily seem to get in the way."

a. Basic empathy.

_____

b. Your hunch and your reason for it.

_____

_____

_____

c. Advanced empathy.

_____

_____

_____

4. A man, 40, is talking to a marriage counselor. This is the third time he has come to see the counselor over the past four years. His wife has never come with him. The other times he spent only a session or two with the counselor and then dropped out. In this session he has been talking a great deal about his latest annoyances with his wife. He says: "I could go on telling you what she does and doesn't do. It's a litany. She really knows how to punish, not only me but others. It's her specialty. I don't even know why I keep putting up with it. I want her to come to counseling, but she won't come. So, here I am again, in her place."

a. Basic empathy.

_____

_____

b. Your hunch and your reason for it.

_____

_____

_____

c. Advanced empathy.

_____

_____

_____

5. A nun, 44, a member of a counselor training group, has been talking about her dissatisfaction with her present job. Although a nurse, she is presently teaching in a primary school because, she says, of the "urgent needs" of that school. When pressed, she refers briefly to a history of job dissatisfaction. In the group she has shown herself to be an active, intelligent, and caring woman who tends to speak and act in self-effacing ways. She mentions how obedience has been stressed throughout her years in the religious order. She does mention, however, that things have been "letting up a bit" in recent years. The younger sisters don't seem to be as preoccupied with obedience as she is. She says: "The reason I'm talking about my job is that I don't want to become a counselor and then discover it's another job I'm dissatisfied with. It would be unfair to the people I'd be working with and unfair to my religious order, which is paying for my education. Of course, I have no iron-clad assurance that I'll be put in a job that will enable me to use my counselor training."

The following are supplemental exercises.
They should be filled in only if the you have not yet gotten the point or
feel that further written practice will help you master this part of the helping model.

A. A high-school senior is talking to a school counselor about college and what kinds of courses she might take there. However, she also mentions, somewhat tentatively, her disappointment in not being chosen as valedictorian of her class. She and almost everyone else had expected her to be chosen. She says: "I know that I would have liked to have been the class valedictorian, but I'm not so sure that you are supposed to count on anything like that. They chose Jane. She'll be good. She speaks well and she's very popular. But no one has a right to be valedictorian. I'd be kidding myself if I thought differently. I've done better in school than Jane, but I'm not as outgoing or popular."

a. Basic empathy.

_____

_____

b. Your hunch and your reason for it.

_____

_____

_____

c. Advanced empathy.

_____

_____

_____

B. A college professor, 43, is talking to a friend, who happens to be a counselor, about his values. He is vaguely dissatisfied with his priorities, but has never done much about examining his current values in any serious way. From time to time the two of them talk about values, but no conclusions are reached. He is not married. Work seems to be a primary value. He says: "Well, it's no news to you that I work a lot. There's literally no day I get up and say to myself, 'Well, today is a day off and I can just do what I want.' It sounds terrible when I put it that way. I've been going on like that for about ten years now. It seems that I should do something about it. But it's obviously my choice. I'm doing what I doing freely. No one's got a gun to my head."

a. Basic empathy.

_____

_____

b. Your hunch and your reason for it.

_____

_____

_____

c. Advanced empathy.

_____

_____

_____

C. A divorced woman, 35, with a daughter, 12, is talking to a counselor about her current relationship with men. She mentions that she has lied to her daughter about her sex life. She has told her that she doesn't have sexual relations with men, but she does. In general she seems quite protective of her daughter. From what her mother says, however, the young girl does not seem to have any serious problems. She says: "I don't want to hurt my daughter by letting her see my darker side. I don't know whether she could handle it. What do you think? I'd like to be honest and tell her everything. I just don't want her to think less of me. I like sex. I've been used to it in marriage, and it's just too hard to give it up. I wish you could tell me what to do about my daughter. I want one of those relationships where we trust each other completely."

a. Basic empathy.

_____

_____

b. Your hunch and your reason for it.

_____

_____

_____

c. Advanced empathy.

_____

_____

_____

INFORMATION SHARING: FROM NEW PERSPECTIVES TO ACTION

As noted in the text, sometimes clients do not get a clear picture of a problem situation or manage it effectively because they lack information needed for clarity and action. Information can provide clients with some of the new perspectives they need to see problem situations as manageable. Giving clients problem-clarifying information or, better, helping them find it themselves is not, of course, the same as advice giving or preaching.

EXERCISE 30: INFORMATION LEADING TO NEW PERSPECTIVES AND ACTION

In this exercise you are asked to consider what kind of information you might give or help clients obtain which would help them see the problem situations they are facing more clearly. Information can, somewhat artificially, be divided into two kinds: (a) information that helps clients understand their difficulties better, and (b) information about what actions they might take.

Example: Tim was a bright, personable young man. During college he was hospitalized after taking a drug overdose during a bout of depression. He spent six months as an in-patient. He was assigned to "milieu therapy," an amorphous mixture of work and recreation designed more to keep patients busy than to help them grapple with their problems and engage in constructive change. He was given drugs for his depression, seen occasionally by a psychiatrist, and assigned to a therapy group that proved to be quite aimless. After leaving the hospital, his confidence shattered, he left college and got involved with a variety of low-paying, part-time jobs. He finally finished college by going to night school, but he avoided full-time jobs for fear of being asked about his past. Buried inside him was the thought, "I have this terrible secret that I have to keep from everyone." A friend talked him into taking a college-sponsored communication-skills course one summer. The psychologist running the program, noting Tim's rather substantial natural talents together with his self-effacing ways, remarked to him one day, "I wonder what kind of ambitions you have." In an instant Tim realized that he had buried all thoughts of ambition. After all, he didn't "deserve" to be ambitious. Tim, instinctively trusting him, divulged the "terrible secret" about his hospitalization for the first time.

90

Tim and the counselor, Rick, had a number of meetings over the course of a year. In some of the early sessions Rick provided him with information that proved challenging in a number of ways. For instance, when it came to employment, the counselor helped Tim find out that:

- he was more intelligent and talented than he realized and, therefore, that he was underemployed;
- that his hospitalization was quite different from, say, a felony conviction;
- deep background checks were not required for the kinds of jobs that Tim might apply for;
- privacy laws often protect prospective employees from damaging disclosures;
- enlightened employers look benignly on the developmental crises A prospect's adolescence;
- the job market was strong and growing stronger so that people with Tim's skills and potential were in high demand.

All this challenged Tim to look at himself in a different light, that is, as a potential winner instead of a loser, and to move more aggressively into job hunting.

In the following cases you are not asked to be an information expert. Rather, from a lay person's or a common-sense point of view, you are asked to indicate what kinds of information you believe might help the client understand and manage the problem situation more clearly. Late specific expert information could be acquired. What blind spots do you think the client might have? How do you think information might help? In what ways would the information be challenging?

1. Point out the blind spots the client might have.
2. Indicate what information could help the client get some new perspectives and move to action.

1. A man, 26, has just been sentenced to five years in a penitentiary on a felony charge. He is talking to a chaplain-counselor who has worked for the past ten years at the penitentiary to which the man has been assigned. The chaplain also counseled the man during his trial. The man is somewhat cavalier about spending time in the penitentiary. He tends to focus on parole.

What blind spots might this client have?

_____

_____

What information might help this client develop new perspectives?

_____

_____

_____

_____

2. A woman, 45, has found a lump in her breast. She is terrified that she has cancer—almost to the point of inaction. She finally talks to a friend of hers who has had breast cancer and is now a member of a self-help group composed of women who have had a mastectomy. She divulges her fears—pain, disfigurement, rejection, death.

What blind spots might this client have?

_____

_____

What information might help this client develop new perspectives?

_____

_____

_____

_____

3. A woman, 28, has been raped and is talking to a counselor at a counseling center for victims of rape. The rape took place two days ago. She has not yet reported the rape to the police. She is terrified and says that she wants to make as little fuss as possible. She feels demeaned and guilty. She fears any involvement with the police. Her instinct is to tell no one in her family.

What blind spots might this client have?

_____

_____

What information might help this client develop new perspectives?

_____

_____

_____

_____

4. A man, 34, comes for counseling because he fears his drinking might be getting out of hand. He's been drinking heavily for several years and recently has had some physical symptoms that he hasn't experienced before, for instance, blackouts. He has managed to remain a secret drinker. At his job he works by himself and has not had more "sick" days than the average. His bouts with drinking tend to be confined to weekends. None of his immediate family or friends has been an alcoholic.

What blind spots might this client have?

_____

_____

What information might help this client develop new perspectives?

_____

_____

_____

_____

5. Tim, 18, has been smoking marijuana for about three years. He is a fairly heavy user. He has recently received several shocks. His father died suddenly and his steady girlfriend left him. Since he was not especially close to his father, he is surprised by how hard he is hit by the loss. After reading a couple of articles on marijuana use, he developed fears that he has been doing irreversible genetic damage to himself. His is fearful of giving it up because he thinks he needs it to carry him over this period of special stress and because he fears withdrawal symptoms.

What blind spots might this client have?

_____

_____

What information might help this client develop new perspectives?

_____

_____

_____

_____

6. Edna, 18, is an unmarried woman who is experiencing her third unexpected and unwanted pregnancy. The other two ended in abortion. She feels guilty about the third unwanted pregnancy and about the abortions. Many of her relatives belong to a fundamentalist church. She is thinking about keeping this child. For all her promiscuity, she seems to know little about sex. She believes that men just take advantage of her.

What blind spots might this client have?

_____

_____

What information might help this client develop new perspectives?

_____

_____

_____

_____

EXERCISE 31: CHALLENGING NORMATIVE DISHONESTIES IN ONE'S OWN LIFE

As indicated in Chapter Ten of the text, most of us face a variety of self-defeating dishonesties in our lives besides the discrepancies. We all allow ourselves, to a greater or lesser extent, to become victims of our own prejudices, smoke screens, distortions, and self-deceptions. In this exercise you are asked to confront some of these, especially the kind of discrepancies that might affect the quality of your helping or the quality of your membership in the training group.

Example 1

- **The issue.** This trainee confronts his need to control situations.
- **The description.** "I am very controlling in my relationships with others. For instance, in social situations I manipulate people into doing what I want to do. I do this as subtly as possible. I find out what everyone wants to do and then I use one against the other and gentle persuasion to steer people in the direction in which I want to go. In the training sessions I try to get people to talk about problems that are of interest to me. I even use empathy and probes to steer people in directions I might find interesting. All this is so much a part of my style that usually I don't even notice it. I see this as selfish, but yet I experience little guilt about it."

Example 2

- **The issue.** This trainee confronts her need for approval from others.

- **The description.** "Most people see me as a 'nice' person. Part of this I like, part of it is a smoke screen. Being nice is my best defense against harshness and criticism from others. I'm cooperative. I compliment others easily. I'm not cynical or sarcastic. I've gotten to enjoy this kind of being 'nice.' I find it rewarding. But it also means that I seldom talk about ideas that might offend others. My feedback to others in the group is almost always positive. I let others give feedback on mistakes. Outside the group I steer clear of controversial conversations. But I'm beginning to feel very bland."

Now confront two of your own "dishonesties," which, if dealt with, will help you be a more effective trainee and counselor. In the description section be as specific as you can. Describe specific experiences, behaviors, feelings. Give brief examples.

a. Issue # 1.

_____

_____

b. Descriptive self-challenge.

_____

_____

_____

_____

_____

a. Issue # 2.

_____

_____

b. Descriptive self-challenge.

_____

_____

_____

_____

_____

_____

_____

HELPER SELF-DISCLOSURE

Although helpers should be **ready** to make disclosures about themselves that would help their clients understand their problem situations more clearly, they should do so only if such disclosures do not upset their clients or distract their clients from the work they are doing. Read the text on helper self-disclosure before doing this exercise.

EXERCISE 32: EXPERIENCES OF MINE THAT MIGHT BE HELPFUL TO OTHERS

In this exercise you are asked to review some problems in living or unused opportunities that you feel you have managed or are managing successfully. Indicate what you might share about yourself that would help a client with a similar problem situation. That is, what might you share of yourself that would help the client move forward in the problem-managing process?

Example 1: "In the past I have been an expert in feeling sorry for myself whenever I had to face any kind of difficulty. I know very well the rewards of seeing myself as victim. I used to fantasize myself as victim as a form of daydreaming or recreation. I think many clients get mired down in their problems because they let themselves feel sorry for themselves the way I did. I think I can spot this tendency in others. When I see this happening, I think I could share brief examples from my own experience and then ask clients to see if what I was doing squares with what they see themselves doing now."

Example 2: "I have been addicted to a number of things in my life and I see a common pattern in different kinds of addiction. For instance, I have been addicted to alcohol, to cigarettes, and to sleeping pills. I have also been addicted to people. By this I mean that at times in my life I have been a very dependent person and I found the same kind of symptoms in dependency that I did in addiction. I know a lot about the fear of letting go and the pain of withdrawal. I think I could share some of this in ways that would not accuse or frighten clients or distract them from their own concerns."

1. List two areas in which you feel you have something to share that might help clients who have problems in living similar to your own. Just briefly indicate the area.
2. Then indicate what you might share.

Area # 1.

_____

_____

What in this area might you share, if appropriate?

_____

_____

_____

_____

_____

Area # 2.

_____

_____

What in this area might you share, if appropriate?

_____

_____

_____

_____

_____

3. Share these with a learning partner. Give and get feedback on the usefulness of the disclosures. Also discuss the conditions underlying appropriateness.

IMMEDIACY: EXPLORING RELATIONSHIPS

As noted in the text, your ability to deal directly with what is happening between you and your clients in the helping sessions themselves is an important skill. **Relationship** immediacy refers to your ability to review the history and present status of your relationship with other members of your group—who are both your fellow trainees and your "clients"—in concrete behavioral ways. **Here-and-now** immediacy refers to your ability to deal with a particular situation that is affecting the ways in which you and another person are relating right now, in this moment.

Immediacy is a complex skill. It involves: (1) revealing how you are being affected by the other person, (2) exploring your own behavior toward the other person, (3) sharing hunches about his or her behavior toward you or pointing out discrepancies, distortions, smoke screens, and the like, and (4) inviting the other person to explore the relationship with a view to developing a better working relationship. For instance, if you see that a client is manifesting hostility toward you in subtle, hard-to-get-

97

at ways, you may: (1) let the client know how you are being affected by what is happening in the relationship (that is, you share your experience), (2) explore how you might be contributing to the difficulty, (3) describe the client's behavior and share reasonable hunches about what is happening (challenge), and (4) invite the client to examine in a direct way what is happening in the relationship. Immediacy involves collaborative problem solving with respect to the relationship itself.

EXERCISE 33: RESPONDING TO SITUATIONS CALLING FOR IMMEDIACY

In this exercise a number of client-helper situations calling for some kind of immediacy on your part are described. You are asked to consider each situation and respond with some statement of immediacy. Consider the following example.

Example

- **Situation.** This client, a man of 44, occasionally makes snide remarks either about the helping profession itself or some of the things that you do in your interactions with him. At times he is cooperative, but at times he asks you to make decisions for him. "Just tell me the best way for me to tell my boss that he's an idiot." He makes remarks about you personally at times, "I bet you've got a lot of friends." Or, on being invited to challenge himself by you, "I hope you're a little more caring with your friends."
- **Immediacy response.** "Tom, let's stop a minute and explore what's happening between you and me in our sessions. . . . (Tom says, "Uh, oh, here we go!") . . . You take mild swipes at the counseling profession such as, 'I hear people are still trying to find out whether counseling works.' Or at me, like, 'Oh, oh, Mr. Counselor is getting a little hot under the collar.' I've ignored these remarks, but in ignoring them I've become a kind of accomplice. Sometimes we seem to be working like a team. Other times you ask me to make decisions for you, you know, the 'just-tell-me' approach. I've let myself get on edge with you and that's not helping us at all. . . . (Tom says, "You want to call my game, huh?") . . . Tom, I guess that I prefer that we not even play games with each other. Perhaps we could spend a little time re-starting our relationship."

1. Critique the approach this counselor takes. What would you change?
2. Share your critique with a learning partner.
3. Consider the following situations and write out an immediacy response that helps the client challenge unhelpful perspectives and actions.
4. Share your responses with a learning partner and give feedback to each other. Together come up with a more effective immediacy response.

1. **The situation.** The client is a person of the opposite sex. You have had several sessions with this person. It has become evident that the person is attracted to you and has begun to make thinly disguised overtures for more intimacy. The person finds you both socially and sexually attractive. Some of the overtures have mild sexual overtones.

Immediacy response.

_____

_____

_____

_____

_____

_____

_____

2. **The situation.** In the first session you and the client, a relatively successful businessman, 40, have discussed the issue of fees. At that time you mentioned that it is difficult for you to talk about money, but you finally settled on a fee at the modest end of the going rates. He told you that he thought that the fee was "more than fair." However, during the last few sessions he has been dropping hints about how expensive this venture is proving to be. He talks about getting finished as quickly as possible and intimates that is your responsibility. You, who thought that the money issue had been resolved, find it still very much alive.

Immediacy response.

_____

_____

_____

_____

_____

_____

3. **The situation.** The client is a male, 22, who is obliged to see you as part of being put on probation for a crime he committed. He is cooperative for a session or two and then becomes quite resistant. His resistance takes the form of both subtle and not too subtle questioning of your competence, questioning the value of this kind of helping, coming late for sessions, and generally treating you like an unnecessary burden.

Immediacy response.

_____

_____

_____

_____

_____

_____

_____

4. The situation. You are a woman. The client, 19, reminds you of your own son, 17, toward whom you have mixed feelings as he struggles to establish some kind of reasonable independence from you. The client at times acts in very dependent ways toward you, telling you that he is glad that you are helping him, asking your advice, and in various ways taking a "little boy" posture toward you. At other times he seems to wish that he didn't have anything to do with you at all and accuses you of being "like his mother."

Immediacy response.

_____

_____

_____

_____

_____

_____

EXERCISE 34:
IMMEDIACY WITH THE OTHER MEMBERS OF YOUR TRAINING GROUP

This exercise makes sense only if you are a member of an experiential training group.

1.   If the group is large, divide up into subgroups of about four trainees per group.
2.   Read the example below.
3.   On separate paper, write out a statement of immediacy for the members of your group. Imagine yourself in a face-to-face situation with each member successively. Deal with real issues that pertain to the training sessions, interactional style, and so forth.
4.   In a round robin, share with each of the other members of the group the statement you have written for him or her. Obviously it would be better to speak it than read it.
5.   The person listening to the immediacy statement should reply with empathy, making sure that he or she has heard the statement correctly.
6.   Then listen to the immediacy statement the other person has for you and then reply with empathy.

7. Finally, discuss for a few minutes the implications of the two statements.
8. Continue with the round robin until each person has had the opportunity to share an immediacy statement with every other member.

Example: Trainee A to Trainee B.

> I notice that you and I have relatively little interaction in the group. You give me little feedback; I give you little feedback. It's almost as if there is some kind of conspiracy of non-interaction between us. I appreciate the way you participate in the group. For instance, the way you invite others to challenge themselves. You do it carefully but without any apology. I think I refrain from giving you feedback, at least negative feedback, because I don't want to alienate you. I do little to make contact with you. I have a hunch that you'd like to talk to me more than you do, but it's just a hunch. I'd like to hear your side of our story.

With a learning partner, identify the elements of immediacy (self-disclosure, challenge, invitation) in this example. Then move on to the exercise.

Section 11
STEP I-B: III. THE WISDOM OF CHALLENGING

The way in which you challenge clients is extremely important. You need to challenge others in such a way that they will *respond* rather than *react* to your invitations. Before doing the following exercise, read Chapter Eleven in *The Skilled Helper*.

EXERCISE 35: EFFECTIVE VERSUS INEFFECTIVE CHALLENGE

In this exercise you are asked to practice on yourself. It goes without saying that you should not practice ineffective challenging on others, even colleagues in your training group.

1. Choose in area in which you feel that you need to challenge yourself. It can relate either to a problem situation or some unused resource or opportunity.
2. Write out a self-challenge statement in which you violate the principles outlined in Chapter Eleven.
3. Then write a self-challenge statement that embodies these principles.
4. Share both forms of self-challenge with a learning partner and discuss. Point out the flaws in the poor self-challenge. Provide feedback to each other on the quality of the appropriate self-challenge.

Example.

• **Area needing challenge**: a certain lack of discipline in my life
• **Poor self-challenge**: You're disorganized and lazy. Your room is continually a mess and your excuse is that you are so busy. But that's a lame excuse. Your wait until the end of the course to do the required papers and therefore then quality is poor. In some ways it is worse in your interpersonal life. You're always late. Others have to wait for you. It's a way of saying you're more important than others. You're just inconsiderate. All of this is so ingrained in you, you're not going to change—unless something drastic happens.

- **Effective self-challenge**:
  Self-as-challenger: You do reasonably well in school and your social life is quite decent. There is the possibility that both could be even more satisfying. For instance, with respect to school. You tend to put things off, sometimes to the point that rushing affects the quality of your work. For instance, end-of-semester papers. Since you pride yourself on giving your best, putting things off like that is not fair to yourself.
  Self-as-responder: Yes, I keep telling myself it's time to get better organized, but I have no plan for doing so. I haven't looked it from the viewpoint of legitimate pride and being fair to myself. That's helpful.
  Self-as-challenger: Let's take a look at your social life. Probably more often than you like, you're late for meetings and appointments.
  Self-as-responder: Absolutely. I feel terrible when I do that, but I do keep doing it.
  Self-as-challenger: Yes, I can almost hear the conversations you have with yourself. But let's take a look at it from the other's point of view. Some laugh. They end of not expecting you to be on time. And others who see being on time as a sign of respect feel put down, almost as if you did not see them as important. I know that's not your attitude, but some read your behavior rather than your heart.

Notice that in this example the challenge takes place through dialogue (even though it is dialogue with oneself). While your effective self-challenge does not have to be done through dialogue (after all, you are talking with yourself), if this helps reinforce the message that challenge is an invitation and that it should be a monologue, then use dialogue.

Now, on separate paper, write out the two different approaches to challenge—one poor and one appropriate—and then follow the instructions above.

As you practice the helping model with your colleagues, you will inevitably use various kinds of challenge. Therefore, when you debrief and get feedback on your interactions, make sure that you get and give feedback on the quality of your challenges.

EXERCISE 36: CHALLENGING ONE'S OWN STRENGTHS

Challenging unused or partially used strengths rather than weaknesses is one of the principles outlined in Chapter Eleven. In this exercise you are asked to confront yourself with respect to your own unused or underused strengths and resources.

1. Briefly identify some problem situations.
2. As in the example, indicate how some unused or underused strength or strengths you have could be applied to the management of each problem situation.
3. In addition, indicate some actions you might take to make use of underused strengths.

Example 1.

- **Problem situation**: "My social life is not nearly as full as I would like it to be."
- **Description of underused strengths or resources**. "I have problem-solving skills, but I don't apply them to the practical problems of everyday life such as my less than adequate social life. Instead of defining goals for myself (making acquaintances, developing friendships) and then seeing how many different ways I could go about achieving these goals, I wait around to see if something will happen to make my social life fuller. I remain passive even though I have the skills to become active."

- **Actions I can take to develop and use resource**: "I can get with a friend who has the same problem and together we can used the problem-management process to take explicit steps to develop a fuller social life. I'm good a brainstorming. I can use that skill here."

Example 2.

- **Problem situation**: Katrina, a trainee in her early 20s, is concerned about bouts of anxiety. She has led a rather sheltered life and now realizes that she needs to "break out" in a number of ways if she is to be an effective counselor. The counseling program has put her in touch with all sorts of people, and this is anxiety provoking. She lists some of the strengths she has that can help her overcome her anxiety.
- **Description of underused strengths or resources and some possible actions**:
  - I'm bright. I know that a great deal of my anxiety comes from fear of the unknown.
  - I'm intellectually adventuresome. I pursue new ideas. Perhaps this can help me be more adventuresome in seeking new experiences. New ideas don't kill me. I bet new experiences won't either.
  - People find me easy to talk to. This can help me form friendships. My fears have kept me from developing friendships. However, developing more relationships is the royal route to the kinds of experiences I need.
  - At root, I'm a religious person. This can help me put my fears into a larger context. In the larger context they seem petty.

Now apply the same procedure to two of your concerns.

a. Problem situation # 1.

_____

_____

_____

Underused strengths as applied to the problem situation and actions to be taken.

_____

_____

_____

Actions you could take to make use of this underused resource in this problem situation.

_____

_____

_____

Problem situation # 2.

_____

_____

_____

Underused strengths as applied to the problem situation and actions to be taken.

_____

_____

_____

_____

Actions you could take to make use of this underused resource in this problem situation.

_____

_____

_____

EXERCISE 37: TENTATIVENESS IN THE USE OF CHALLENGING SKILLS

As noted in the text, challenges are usually more effective if they do not sound like accusations. Therefore, in challenging clients, don't accuse them, but don't be so tentative that the force of the challenge is lost.

1. Return to Exercise 28 on advanced empathy in Section 10.
2. Review your responses with a learning partner in terms of tentativeness.
3. Redo responses that would benefit from more or less or better expressed tentativeness.

EXERCISE 38: THE CHALLENGE ROUND ROBIN

The purpose of this exercise is to give you the opportunity to practice both challenging others and responding non-defensively and creatively to those who challenge you. The assumption is that you have begun to know the other members of your training group fairly well and that you are thoroughly familiar with the principles and methodology of effective challenging.

1. Review the material on challenging and effective response to challenge in Chapters Eight and Nine of the text.

2. Divide up into groups of three. The three roles are challenger, person being challenged, and observer. Challengers will do two things: (a) point out something they have noticed that their partners do well in the training group and then (b) challenge their partners in some way (for instance, by pointing out a strength or resource that is being underused or some other discrepancy, being careful to be descriptive rather than accusatory).

3. The person being challenged first responds with basic empathy to make sure that he or she understands the point of the challenge.

4. At this point, the observer gives feedback both to the challenger in terms of accuracy, usefulness, and manner or style. The observer also gives feedback to the person being challenged on how well he or she responded to the challenge.

5. Then the partners, using this feedback to the degree that it is helpful, briefly explore the area challenged in terms of concrete experiences, behaviors, and feelings.

6. Each person in the triad is given a chance to play all three roles.

Example

Challenger. "In our group sessions, you take pains to see to it that other members of the group are understood, especially when they talk about sensitive issues. You provide a great deal of empathy and you encourage others, principally by your example, to do the same. Your empathy never sounds phony and you're usually quite accurate.

"I've also noticed that you tend to limit yourself to basic empathy. You seldom use probes and you seem to be slow to challenge anyone, for instance, by sharing hunches about the issues they are discussing. Because of your empathy and your genuineness, you have amassed a lot of 'credits' in the group, but you don't use them to help others make reasonable demands on themselves."

Person challenged. "You appreciate my ability and willingness to be empathic. But I am less effective than I might be in that I don't move beyond empathy, especially since I 'merit' doing so. I should work on increasing my challenging skills. In effect my 'range' within the group is too narrow. If I were to expand the kinds of responses I use, I'd be more helpful."

At this point the observer provides feedback and then A and B then spend a few minutes exploring the issue that has been raised. Rotate so that each person gets to play each the role.

Section 12
STEP I-C: LEVERAGE
Helping Clients Work on the Right Things

The exercises in this section deal with helping clients work on issues that will make a difference in their lives, that is, issues that have "leverage." Counseling time is too precious to waste on issues that do not add value and make a difference in the client's life. The final exercise in this section helps you pull together all you have learned about Stage I of the helping process. Read Chapter Twelve in *The Skilled Helper* before doing the exercises inthis section.

EXERCISE 39: MYSELF AS A DECISION MAKER

Since, as a counselor, you are going to be helping clients make decisions (which, of course, is not the same as making decisions for them), it is useful to take a look at yourself as decision maker.

1. Read the parts of the text dealing with decision making.
2. With a learning partner, discuss the difference between "rational" decision making and the "shadow side" of decision making.
3. On your own, review the way you go about making decisions. Then, on a separate piece of paper, write a short description of your decision-making style. Include the "shadow side" of your decision-making style.
4. Swap descriptions with a learning partner. Read them. Then discuss on you might improve your style.

EXERCISE 40: SCREENING PROBLEMS

Clients may need help in determining whether their issues are important enough to bring to a helper in the first place. This is called "screening."

1. Read these two case summaries with a view to discussing them with your fellow trainees.

- **Case 1**: Lila comes to a counselor for help. In telling her story, she says that she has occasional headaches and a few arguments with her husband, Lance. Doctors have told her that there is nothing wrong with her physically. The helper, using empathy, probing, and an occasional challenge, discovers that Lila is not understating the nature of her concerns and that there are no further hidden issues. She does discover that Lila and Lance have no children, that Lance works rather long hours in his new job, and that Lila is mainly a householder in their small condominium and has few outside interests.

- **Case 2**: Ray, 41, is a middle manager in a manufacturing company located in a large city. He goes to see a counselor with a somewhat complex story. He is bored with his job; his marriage is lifeless; he has poor rapport with his two teenage children, one of whom is having trouble with drugs; he is drinking heavily; his self-esteem is low; he has begun to steal things, small things, not because he needs them but because he gets a kick out of it. He tells his story in a rather disjointed way, skipping around from one problem area to another. He is a talented, personable, engaging man who seems to be adrift in life. He does not show any symptoms of severe psychiatric illness. He does experience a great deal of uneasiness in talking about himself. This is his first visit to a helper.

2. Discuss these two cases in terms of the material in the text on screening. Consider these questions:

- How would you approach the woman in Case 1?
- Under what conditions would you be willing to work with her?
- How would you approach the man in Case 2?
- In what general ways does this case differ from Case 1?
- What would your concerns be in working with the man in Case 2?

EXERCISE 41: CHOOSING ISSUES THAT MAKE A DIFFERENCE

Often enough, the stories clients tell are quite complex. And so they may need your help in deciding which issues to work on first and which merit substantial attention. In this exercise you are asked to review the personal concerns and problems you have identified in doing the exercises in this manual.

1. Briefly list about ten concerns or unexploited opportunities you have discovered in doing the assessment exercises up to this point.

_____

_____

_____

_____

_____

_____

_____

_____

2. Now do some screening. Put a line through those that you would probably not bring to a counselor because they are not that important or because you believe that you could handle them easily if you wanted.

3. Review the rest in the light of the following "leverage" criteria taken from the text.

- a. If there is a crisis, first help the client manage the crisis.
- b. Begin with the problem that seems to be causing pain for the client.
- c. Begin with issues the client sees as important and is motivated to work on.
- d. Begin with some manageable sub-problem of a larger problem situation.
- e. Begin with a problem that, if handled, will lead to some kind of general improvement in the client's condition.
- f. Focus on a problem for which the benefits will outweigh the costs.

4. Finally, using the criteria below, evaluate the items remaining on your list. Next to each concern place the letters of the applicable "leverage" criteria listed below.

Example: Gino is a trainee in a clinical psychology program. Here is one of the concerns on his list: "I am very inconsistent in the way I deal with people. For instance, some of my friends see me as fickle, I blow hot and cold. One friend told me that whenever he sees me, he's not sure which Gino he will meet. In fact, I seem to be inconsistent in other areas of life. Sometimes I give myself wholeheartedly to my studies, sometimes I couldn't care less." Gino believes that the following criteria apply: c, e, f.

5. Choose two issues that have high-leverage value for you. These are issues, concerns, opportunities, or problems, that, if pursued, would make a difference in your life.

6. Explain, in terms of the criteria outlined above or any further criteria not listed there, why you think that each has leverage. What is the "bang for the buck" of each?

7. Share one of these with a learning partner. Discuss what it would take to get you to invest time and energy in working with this problem and/or opportunity.

PULLING STAGE I TOGETHER

The following exercise asks you to pull together what you have learned about Stage I of the helping process and apply it to yourself.

EXERCISE 42: COUNSELING YOURSELF: AN EXERCISE IN STAGE I

At this stage, you have developed an overview of the helping model and, through reading and doing the exercises in this manual, developed an understanding and behavioral "feel" for Stage I. In this exercise you are asked to carry on a dialogue with yourself in writing. Choose a problematic area of your life, one that is relevant to your hoped-for success as a helper. First use empathy, probes, and challenges to help yourself tell your story. Choose a high-leverage issue for exploration. Work at clarifying it in terms of specific experiences, behaviors, and feelings. The dialogue should include probes for and challenges to engage in problem-managing and opportunity-developing action.

Example: This example comes from the experience of Cormack, a man in a master's degree program in counseling psychology. Here, then, is Cormack's dialogue with himself.

His Initial Story. "To be frank, I have a number of misgivings about becoming a counselor. A number of things are turning me off. For instance, one of my instructors this past semester was an arrogant guy. I kept saying to myself, 'Is this what these psychology programs produce? Could this guy really help anyone?' I also find the program much too theoretical. In a 'Theories of Counseling and Psychotherapy' course we never did anything, not even discuss the practical implications of these theories. And so they remained just that—theories. I'm very disappointed. I'm about to go into my second year, but I've got serious reservations. From what others tell me, the program gets a bit more practical, but not enough. There's a practicum experience at the end of the program, but I need more hands-on work now. So I've started working at a halfway house for people discharged from mental hospitals. But that's not working out the way I expected either. There's something about this whole helping business that is making me think twice about myself and about the profession."

RESPONSE TO SELF: All of this adds up to the fact that the helping profession, at least from your experience, is not what it's cracked up to be. The program and some of the instructors in it have left you quite disappointed. Of all these issues, which one or ones hits you the hardest?

SELF: Hmm. It's hard to say, but I think the halfway house bothers me most. Because that's not theoretical stuff. That's real stuff out there.

RESPONSE TO SELF: That's a place where real helping should be taking place. But you've got misgivings about what's going on there.

SELF: Yes, two sets of misgivings. One set about me and one about the place.

RESPONSE TO SELF: Which set do you want to explore?

SELF: I feel I have to explore both, but I'll start with myself. I feel so ill-prepared. What's in the lectures and books seems so distant from the realities of the halfway house. For instance, the other day one of the residents there began yelling at me when we were passing in the hallway. She hit me a few times and then ran off screaming that I was after her. I felt so incompetent. I felt guilty.

RESPONSE TO SELF: That sounds pretty upsetting. You just weren't prepared for it. I'm wondering whether the more practical part of the counseling program you're going into would better prepare you for that kind of reality.

SELF: It could be. I may be doing myself in by jumping ahead of myself.

RESPONSE TO SELF: But you still have reservations about the effectiveness of the halfway house.

SELF: I wasn't ready for what I found there. I've been there a couple of months. No one has really helped me learn the ropes. I don't have an official supervisor. I see all sorts of people with problems and help when I can.

RESPONSE TO SELF: You just don't feel prepared and they don't do much to help you. So you feel inadequate. You also seem to be basing your judgments about the adequacy of helper-training programs and helping facilities on this training program and on the halfway house.

SELF: That's a good point. I'm making the assumption that both should be high-quality places. As far as I can tell, they're not. I guess I have to work on myself first. But that's why I went to the halfway house in the first place. I'm an independent person, but I'm too much on my own there. In a sense, I'm trusted, but, since I don't get much supervision, I have to go on my own instincts and I'm not sure they're always right.

RESPONSE TO SELF: There's some comfort in being trusted, but without supervision you still have a what-am-I-doing-here feeling.

SELF: There are many times when I ask myself just that, 'What are you doing here?' I provide day-to-day services for a lot of people. I listen to them. I take them places, like to the doctor. I get them to participate in conversations or games and things like that. But it seems that I'm always just meeting the needs of the moment. I'm not sure what the long-range goals of the place are and if anyone, including me, is contributing to them in any way.

RESPONSE TO SELF: You get some satisfaction in providing the services you do, but this lack of overall purpose or direction for yourself and the institution is frustrating. I'm not sure what you're doing about all of this, either at the halfway house or in the counseling program.

SELF: I'm letting myself get frustrated, irritated, and depressed. I'm down on myself and down on the people who run the house. It's a day-to-day operation that sometimes seems to be a fly-by-night venture. See! There I go. You're right. I'm not doing anything to handle my frustrations. I am doing something to try to better myself, that is, to make myself a better helper, but this is the first time I've expressed myself about the halfway house or about the counseling program.

RESPONSE TO SELF: So this is really the first time you've stopped to take a critical look at yourself as a potential helper and the settings in which helping takes place. And what you see is depressing.

SELF: Absolutely.

RESPONSE TO SELF: I wonder the how fair you're being to yourself, to the profession, and even to the halfway house.

SELF (Pause): Well, that's a point. I'm speaking—and making judgments—out of a great deal of frustration. I don't want to go off half cocked. But I could share my concerns about the psychology program with one of the instructors. She teaches the second-year trainees. She gave a very practical talk. I could also see whether my concerns are shared by my classmates. I could also talk to the second-year students to get a feeling for how practical next year might be.

RESPONSE TO SELF: Give yourself a chance to get a more balanced perspective.

SELF. "Right. I need to do something instead of just brooding and complaining. But there's still the halfway house. I don't want to come across as the wet-behind-the-ears critic.

RESPONSE TO SELF: I'm not sure whether you're assuming that particular halfway house should be a state-of-the-art facility?

SELF: Touche! . . . . Wow, I am basing a lot of my feelings about the profession on that place.

RESPONSE TO SELF: Who out there might you trust to help you get a better perspective?

SELF: No one I can think of. . . . But there is a consultant who shows up once in a while. He runs staff meetings in a very practical way. Maybe I can get hold of him and get a wider picture.

RESPONSE TO SELF: So, overall?

SELF: I've got some work to do before I rush to judgment. I like the fact that I want the people and the institutions in the profession to be competent. Deep down I think that I'd make a good helper. I say to myself, 'You're all right; you're trying to do what is right.' Also, I need to challenge my idealism. I spend too much time grieving over what is happening at school and at the halfway house. I need to figure out how to turn minuses into pluses.

1. Review this trainee's responses to himself with a learning partner. What kinds of responses did he use? How would you evaluate their quality? Did they get him someplace? Describe the movement he made during the session. What responses would you have changed?
2. Choose a problematic area that is important to you and on separate sheets of paper engage in the same kind of dialogue with yourself. Tell your story briefly, choose a high-leverage issue, and clarify it in terms of specific experiences, behaviors, and feelings. Stay within the steps of Stage I.
3. Exchange dialogues with your learning partner and engage in the same kind of critique outlined above.

PART FOUR

STAGE II: HELPING CLIENTS DETERMINE WHAT THEY NEED AND WANT

Part Four, which includes Stages II and III of the helping model, is in many ways the heart of helping. It is here that counselors help clients determine what they need and want and develop programs for constructive change. The payoff for identifying and clarifying both problem situations and unused opportunities lies in doing something about them. The skills that both helpers and clients need to do precisely this are outlined and illustrated in Chapters Thirteen through Eighteen in *The Skilled Helper*.

STAGE II
Helping Clients Create A Better Future

Stage II focuses on a better future, the client's preferred scenario. Problems can make clients feel hemmed in and closed off. To a greater or lesser extent they have no future, or the future they have looks troubled. The steps of Stage II outline three ways in which helpers can be with their clients at the service of exploring and developing this better future.

Step II-A: Helping clients identify *possibilities* for a better future. What do you want? What do you need? What are some of the possibilities?
Step II-B: Helping clients choose specific preferred-scenario *goals*. Given the possibilities, what do you really want? What are your choices?
Step II-C: Helping clients discover incentives for *commitment* to their goals. What are you willing to pay for what you want?

Exercises related to Stage II are found in Sections 13 through 15.

Section 13
STEP II-A: WHAT DO YOU NEED AND WANT?
Possibilities for a Better Future

Effective helping requires imagination. In this step you are asked to help yourself and clients develop a vision of a better future. Once clients understand the nature of the problem situation, they need to ask themselves, "What do I want. What would my situation look like if it were better, at least a little bit better?" Read Chapter Thirteen in the text before doing these exercises.

EXERCISE 43:
PROBLEMS AND OPPORTUNITIES IN THE SOCIAL SETTINGS OF LIFE

This is another exercise that will help you review areas in which clients have problems and also help you identify your own strengths and soft spots. Individuals belong to and participate in a number of different social settings in life: family, circle of friends, clubs, church groups, school groups, and the like. People are also affected by what goes on in our neighborhoods and the cities and towns in which we live. Larger systems, such as state and national governments, have their ways of entering people's lives.

1. **Chart the social settings of life.** In this exercise you are asked to write you name in the middle of a sheet of paper. Then, as in the example (Figure 1), draw spokes out to the various social settings of your life. The person in the example is Mitch, 45, a principal of an inner-city high school in a large city. He is married and has two teenage sons. Neither attends the high school of which he is principal. He is seeing a counselor because of exhaustion and bouts of hostility and depression. He has had a complete physical check-up and there is no evidence of any medical problem.
2. **Choose two or three social settings that are currently key**. These are settings in which you are experiencing problems or have unused opportunities.
3. **Review issues, demands, conflicts, concerns.** Now take each key social setting and write down issues or concerns in that setting. For instance, some of things Mitch writes are:

School

*   Some faculty members want a personal relationship with me and I have neither the time nor the desire.
*   Some faculty members have retired on the job. I don't know what to do with them.
*   Some of the white faculty members are suspicious of me and distant just because I'm black.
*   One faculty member wrote the district superintendent and said that I was undermining her reputation with other faculty members. This is not true.
*   The students, both individually and through their organizations, keep asking me to be more liberal while their parents are asking me to tighten things up.

Family

*   My wife says that I'm letting school consume me; she complains constantly because I don't spend enough time at home. Even though she wants me to spend more time at home and yet she criticizes me for not spending more time with my parents.
*   My kids seem to withdraw from me because I'm a double authority figure, a father and a principal.

Parents

- My mother is infirm; my retired father calls me and tells me what a hard time he's having getting used to retirement.
- My mother tells me not to be spending time with her when I have so much to do and then she complains to my wife and my father when I don't show up.

School District

- I would like to become the district superintendent. I think I could do a lot of good.

Friends

- My friends say that I spend so much time at work involving myself in crisis management that I have no time left for them; they tell me I'm doing myself in.

4. **Possibilities for a better future.** Choose two concerns arising from two different social settings of your life. Spell out some of the possibilities for a better future in this area. What would this problem or unused opportunity would look like if it were managed better or solved? What are some of the things that would be in place that are not now in place? Use the following probes to help yourself brainstorm possibilities.

- Here's what I need . . . .
- Here's what I want . . . .
- Here are some items on my wish list . . . .
- When I'm finished I will have . . . .
- There will be . . . .
- I will have in place . . . .
- I will consistently be . . . .
- There will be more of . . . .
- There will be less of . . . .

For instance, Mitch, in reviewing the conflict between his work and his friends, comes up with these possibilities: more time with his friends, greater flexibility in managing his calendar at work, better integration of his friendships with his home life, a more clear-cut division between his work setting and the other social settings of his life, more time for himself, a clearer understanding of his career aspirations and the costs associated with them, and better delegation on his part to the members of the school administrative team. Now do the same for yourself in two or three of the social settings of your life on a separate sheet of paper.

a. First social setting.

_____

b. Social-setting related issue, demand, conflict, opportunity, concern.

_____

_____

_____

c. Possibilities for a better future in this area of concern.

_____

_____

_____

_____

_____

_____

a. Second social setting.

_____

b. Social-setting related issue, demand, conflict, opportunity, concern.

_____

_____

_____

c. Possibilities for a better future in this area of concern.

_____

_____

_____

_____

_____

_____

a. Third social setting.

_____

b. Social-setting related issue, demand, conflict, opportunity, concern.

_____

_____

_____

c. Possibilities for a better future in this area of concern.

_____

_____

_____

_____

_____

_____

_____

EXERCISE 44: BRAINSTORMING POSSIBILITIES FOR A BETTER FUTURE—CASES

Here area number of cases, most of which you have seen before. Now you are asked to put yourself in each person's shoes and brainstorm possibilities for a better future. What would things look like if they looked better? If you were this person, what are some of the things you would want instead of what you've got?

1. Read the case. Put yourself in this person's shoes. Make his or her problem situation your own. Do not see the exercise as an exercise in advice giving.
2. Point out a blind spot this person might have—a blind spot that might keep this person from managing the problem situation effectively.
3. Brainstorm of number of possibilities for a better future. What would you want if you were this person?
4. Use the probes outlined in the previous exercise to help yourself develop preferred-scenario possibilities:

1. Tim was a bright, personable young man. During college he was hospitalized after taking a drug overdose during a bout of depression. He spent six months as an in-patient. He was assigned to "milieu therapy," an amorphous mixture of work and recreation designed more to keep patients busy than to help

them grapple with their problems and engage in constructive change. He was given drugs for his depression, seen occasionally by a psychiatrist, and assigned to a therapy group that proved to be quite aimless. After leaving the hospital, his confidence shattered, he left college and got involved with a variety of low-paying, part-time jobs. He finally finished college by going to night school, but he avoided full-time jobs for fear of being asked about his past. Buried inside him was the thought, "I have this terrible secret that I have to keep from everyone." A friend talked him into taking a college-sponsored communication-skills course one summer. The psychologist running the program, noting Tim's rather substantial natural talents together with his self-effacing ways, remarked to him one day, "I wonder what kind of ambitions you have." In an instant Tim realized that he had buried all thoughts of ambition. After all, he didn't "deserve" to be ambitious. Tim, instinctively trusting him, divulged the "terrible secret" about his hospitalization for the first time.

a. A key blind spot translated into an future-oriented new perspective.

_____

_____

b. A range of possibilities for a better future.

_____

_____

_____

_____

2. Cormack in a master's degree program in counseling psychology: "To be frank, I have a number of misgivings about becoming a counselor. A number of things are turning me off. For instance, one of my instructors this past semester was an arrogant guy. I kept saying to myself, 'Is this what these psychology programs produce? Could this guy really help anyone?' I also find the program much too theoretical. In a 'Theories of Counseling and Psychotherapy' course we never did anything, not even discuss the practical implications of these theories. And so they remained just that—theories. I'm very disappointed. I'm about to go into my second year, but I've got serious reservations. From what others tell me, the program gets a bit more practical, but not enough. There's a practicum experience at the end of the program, but I need more hands-on work now. So I've started working at a halfway house for people discharged from mental hospitals. But that's not working out the way I expected either. There's something about this whole helping business that is making me think twice about myself and about the profession."

a. A key blind spot translated into an future-oriented new perspective.

_____

_____

b. A range of possibilities for a better future.

_____

_____

_____

_____

_____

3. A nun, 44, a member of a counselor training group, has been talking about her dissatisfaction with her present job. Although a nurse, she is presently teaching in a primary school because, she says, of the "urgent needs" of that school. When pressed, she refers briefly to a history of job dissatisfaction. In the group she has shown herself to be an active, intelligent, and caring woman who tends to speak and act in self-effacing ways. She mentions how obedience has been stressed throughout her years in the religious order. She does mention, however, that things have been "letting up a bit" in recent years. The younger sisters don't seem to be as preoccupied with obedience as she is. She says: "The reason I'm talking about my job is that I don't want to become a counselor and then discover it's another job I'm dissatisfied with. It would be unfair to the people I'd be working with and unfair to my religious order, which is paying for my education. Of course, I have no iron-clad assurance that I'll be put in a job that will enable me to use my counselor training."

a. A key blind spot translated into an future-oriented new perspective.

_____

_____

b. A range of possibilities for a better future.

_____

_____

_____

_____

_____

4. A college professor, 43, is talking to a friend, who happens to be a counselor, about his values. She is vaguely dissatisfied with her priorities, but has never done much about examining her current values in any serious way. From time to time the two of them talk about values, but no conclusions are reached. She is not married. Work seems to be a primary value. She says: "Well, it's no news to you that I work a lot. There's literally no day I get up and say to myself that today is a day off that I can use anyway I want. It sounds terrible when I put it that way. I've been going on like that for about ten years now. It

seems that I should do something about it. But it's obviously my choice. I'm doing what I doing freely. No one's got a gun to my head."

a. A key blind spot translated into an future-oriented new perspective.

_____

_____

b. A range of possibilities for a better future.

_____

_____

_____

_____

5. This man, now 64, retired early from work when he was 62. He and his wife wanted to take full advantage of the "golden" years. But his wife died a year after he retired. At the urging of friends he has finally come to a counselor. He has been exploring some of the problems his retirement has created for him. His two married sons live with their families in other cities. In the counseling sessions he has been dealing somewhat repetitiously with the theme of loss. He says: "I seldom see the kids. I enjoy them and their families a lot when they do come. I get along real well with their wives. But now that my wife is gone . . . (pause) . . . and since I've stopped working . . . (pause) . . . I seem to just ramble around the house aimlessly, which is not like me at all. I suppose I should get rid of the house, but it's filled with a lot of memories—bittersweet memories now. There were a lot of good years here."

a. A key blind spot translated into an future-oriented new perspective.

_____

_____

b. A range of possibilities for a better future.

_____

_____

_____

_____

The following are supplemental exercises.
They should be filled in only if the you have not yet gotten the point or
feel that further written practice will help you master this part of the helping model.

A. A man, 48, husband and father, is exploring the poor relationships he has with his wife and children. In general, he feels that he is the victim, that his family is not treating him right (that is, like many clients, he emphasizes his experience rather than his behavior). He has not yet examined the implications of the ways he behaves toward his family. At this point he is talking about his sense of humor. He says: "For instance, I get a lot of encouragement for being witty at parties. Almost everyone laughs. I think I provide a lot of entertainment, and others like it. It also works on my job. But this is another way I seem to flop at home. When I try to be funny, my wife and kids don't laugh, at least not much. At times they even take my humor wrong and get angry. I actually have to watch my step in my own home."

a. A key blind spot translated into an future-oriented new perspective.

_____

_____

b. A range of possibilities for a better future.

_____

_____

_____

_____

B. A divorced woman, 35, with a daughter, 12, is talking to a counselor about her current relationship with men. She mentions that she has lied to her daughter about her sex life. She has told her that she doesn't have sexual relations with men, but she does. In general she seems quite protective of her daughter. From what her mother says, however, the young girl does not seem to have any serious problems. She says: "I don't want to hurt my daughter by letting her see my darker side. I don't know whether she could handle it. What do you think? I'd like to be honest and tell her everything. I just don't want her to think less of me. I like sex. I've been used to it in marriage, and it's just too hard to give it up. I wish you could tell me what to do about my daughter. I want one of those relationships where we trust each other completely."

a. A key blind spot translated into an future-oriented new perspective.

_____

_____

b. A range of possibilities for a better future.

_____

_____

_____

_____

C. A first-year engineering graduate student has been exploring his disappointment with himself and with his performance in school. His father is a successful engineer, but did not pressure his son into engineering. He has explored such issues as his dislike for the school and for some of the teachers with his counselor. He says: "I just don't have much enthusiasm. My grades are just okay, maybe even a little below par. I know I could do better if I wanted to. I don't know why my disappointment with the school and some of the faculty members can get to me so much. It's not like me. Ever since I can remember—even in primary school, when I didn't have any idea what an engineer was—I've wanted to be an engineer. Theoretically, I should be as happy as a lark because I'm in a graduate school with a good reputation, but I'm not."

a. A key blind spot translated into an future-oriented new perspective.

_____

_____

b. A range of possibilities for a better future.

_____

_____

_____

_____

D. Hannah, a new counselor working in a hospital setting for a number of months, has begun to confide in one of her colleagues who has worked in health-care facilities for about seven years. Hannah has intimated, however tentatively, that she feels that some of the medical doctors tend to dismiss her because she is new, because she is a counselor and not a medical practitioner, and because she is a woman. One doctor has been particularly nasty, even though he masks his abusiveness under his brand of "humor." One day her colleague says to her, "Tell 'em what you think!" Hannah replies: "That will only make them worse. I think it more important to stand up for my profession by just being who I am and helping patients

120

as much as possible. Their meanness is a punishment in itself. Anyway, it's a game that doctors play with everyone else in the hospital. That kind of thing has been going on forever. It's not such a big deal anyway."

a. A key blind spot translated into an future-oriented new perspective.

_____

_____

b. A range of possibilities for a better future.

_____

_____

_____

_____

E. Len, married with three teenage children, lost all the family's savings in a gambling spree. Under a lot of pressure from both his family and his boss, he started attending Gamblers Anonymous meetings. He seemed to recover, that is, he stopped betting on horses and ball games. A couple of years went by and Len stopped going to the GA meetings because he "no longer needed them." He even got a better job and was recovering financially quite well. But his wife began to notice that he was on the phone a great deal with his broker. She confronted him and he reluctantly agreed to a session with one of the friends he made at GA. He says, "She's worried that I'm gambling again. You know, it's really just the opposite. When I was gambling I was financially irresponsible. I lost our future on horses and ball games. But now I'm taking a very active part in creating our financial future. Every financial planner will tell you that investing in the market is central to sound financial planning. I'm a doer. I'm taking a very active role."

a. A key blind spot translated into an future-oriented new perspective.

_____

_____

b. A range of possibilities for a better future.

_____

_____

_____

EXERCISE 45: BECOMING AN EFFECTIVE HELPER

You may or may not want to become either a paraprofessional or professional helper. You may be involved in this course simply to become a better helper in everyday life, not matter what career you choose to pursue. In this exercise assume that one of these three options is yours.

1. Describe what you are currently like as a helper.
2. Describe, in terms of possibilities, the kind of helper you would like to become. Be as concrete as possible.

a. Here is what I am currently like.

_____

_____

_____

_____

_____

b. Here is what I'd like to become.

_____

_____

_____

_____

_____

_____

3. Share your two descriptions with a learning partner. Mutually clarify and refine these possibilities through empathy and probing. After listening to the other, indicate further possibilities you might want to add to your list.

EXERCISE 46:
HELPING ANOTHER DEVELOP POSSIBILITIES FOR A BETTER FUTURE

In this exercise, you are asked to help one of the other members of your group develop possibilities for a better future.

1. Divide up into groups of three for this exercise.
2. Assume, sequentially, the roles of client, helper, and observer.
3. In the role of client, provide a summary of one of the problem situations or undeveloped opportunities focused on in any of the previous exercises. Or, if you prefer, choose a new issue.
4. In the role of counselor, listen to the client summarize the problem situation or undeveloped opportunity. Use empathy and probes to get a full statement of the issue.
5. Finally, help the other develop possibilities for a better future. Use empathy, probes, summaries, and challenges to help the other expand the list.

Example: Geraldo, a junior in college majoring in business studies, has given the following summary of the problem situation: "My uncle runs a small business. He's offered me a kind of internship in the family business. While such an opportunity fits perfectly with my career plans, I have done nothing to develop it. Right now, it's only a good idea. But it still seems to good an opportunity to miss." Trish, his helper, uses all the communication skills learned up to this point to help him develop internship possibilities. Some of the dialogue goes like this:

TRISH. So you haven't taken the time yet to spell out what the internship could look like.
GERALDO. No. Generally, I see the internship as something that would bring to life the stuff I'm getting out of business books right now.
TRISH. So the internship would give a chance for some hands-on experience with business realities that are just concepts now. What could the internship look like?
GERALDO. Well, I don't know what's on my uncle's mind. Though he's not going to push me into anything I don't want.
TRISH. Then it seems to make sense for you to have some ideas in mind before you talk with him about it. What would *you* like the internship to look like? What do you want?
GERALDO. Let's see. I'd be putting in 10 to 15 hours a week at my uncle's business. Actually, that would make sense since I'm more or less overbalanced right now on social activities.
TRISH. What would you be doing during those hours?
GERALDO. Well, I don't like the finance part of my program. So one thing I could do is become familiar with the finance part of his business—where the money comes from, what kind of debt he carries, cash flow—all those things that are still too theoretical for me in the courses I'm taking.
TRISH. You'd see finance in action. It might even be more interesting then. What do you picture yourself doing?
GERALDO: Since it's s small company, I guess I would being dong it by talking to my uncle. I'm sure he takes the lead in financial decisions. I'd be comparing what the books say with what he actually does.
TRISH. Who knows? Maybe some of the latest tools you're learning about would be useful to him. It might go both ways. What else do you see yourself doing?
GERALDO: I've read a lot about corporate culture and the way it can either strengthen or strangle a business. I'd like to see if there's a culture in my uncle's business and what it's like. There are only about a hundred people working there. So it's lab size.

123

TRISH. That would make it more than just work, but a place to learn a lot of practical things. What would you be doing about culture?

GERALDO. Yeah. It would certainly send me back to the books in a different way. I saw a couple of articles on new approaches to strategic management and on employee empowerment in the *Harvard Business Review*. But they were just theory. My uncle's place would help me turn theory into practice.

TRISH. So there could be a lot of synergy between the internship and your studies. Tell me a little more about the corporate culture thing.

GERALDO: Let's see. I could do an informal survey about the beliefs and values and norms that really run the place. I could show it to my uncle and have him see if there are some practical business things in it.

TRISH: All right. What other possibilities are there?

GERALDO. I'd like to learn something about the role of the manager. Especially in directing the work of others. I'm not sure that a manager really does with his day. The texts we have seem to be describing what management used to be, not what it is today. What you read in *Fortune* and *Business Week* seems so different from the texts.

TRISH. So you could find out whether the managers at your uncle's place look more like the text books or the business magazine articles.

GERALDO. Or neither! I have no idea how out-of-date or up-dated my uncle's managers are.

TRISH. What would you be doing to find out?

GERALDO. Well, I could watch them at work. Some managers have people "shadow" them, watch their style and all that and then give them feedback on it so they can do things better.

TRISH. So some form of "shadowing" is a possibility.

GERALDO. I'm not sure my uncle would let me do that.

TRISH. We can figure that out later. Let's keep the possibilities going for the time being.

GERALDO. Right. Well, I could . . . .

The dialogue goes on in that vein. Geraldo, with the help of Trish, develops not only a lot of possibilities for developing the internship opportunity.

6. After ten minutes, the helper gets feedback from the client first and then the observer as to how useful he or she has been.

7. Repeat the process until each person has played each role.

8. Before beginning, the members of the triad should critique the way Trish went about helping Geraldo develop preferred-scenario possibilities. What did she do well? What could have been done better?

Section 14
STEP II-B: WHAT DO YOU REALLY WANT?
Moving from Possibilities to Choices

Once clients have brainstormed possibilities for a better future, they need to make some choices. Answers to the questions, "What do you want" and "What do you need?" are possible goals. Answers to the question, "What do you really need and want," are the actual goals that constitute the change agenda. It is as if the client says, "Here is what I want in place that is not now in place." Goals are a critical part of an agenda for constructive change. For instance, Geraldo, the college student seen in Section 12, after brainstorming a number of possibilities for structuring his internship in his uncle's company, chooses four of the possibilities and makes these his agenda. He shares them with his uncle and, after some negotiation, comes up with a revised package they can both live with. Read Chapter Fourteen before doing these exercises.

EXERCISE 47: TURNING POSSIBILITIES INTO GOALS—A CASE

In this exercise you are asked to do what Geraldo had to do—choose several preferred-scenario possibilities as the first step in crafting an agenda. Possibilities should be chosen because they will best help you manage some problem situation or develop some opportunity. First, read the following case.

The Case. Vanessa, 46, has been divorced for about a year. She has done little to restructure her life and is still in the doldrums. At the urging of a friend, she sees a counselor. With her help, she brainstorms a range of possibilities for a better future around the theme of "my new life as a single person." She is currently a salesperson in the women's apparel department of a moderately upscale store. She lives in the house that was part of the divorce settlement. She has no children. When she finally decided that she wanted children, it was too late. There was some discussion with her ex-husband about adopting a child, but it didn't get very far. The marriage was already disintegrating. Her visits to the counselor have reawakened a desire to take charge of her life and not just let it happen. She let her marriage happen and it fell apart. She is not filled with anger at her former husband. If anything, she's a bit too down on herself. Her grieving is filled with self-recrimination.

The helper asks her, "What do you want now that the divorce is final? What would you like your post-divorce life to look like? What would you want to see in place?" Vanessa brainstormed the following possibilities:

- A job related to the fashion industry.
- A small condo that will not need much maintenance on my part instead of the house.
- The elimination of poor-me attitudes as part of a much more creative outlook on self and life.
- The elimination of waiting around for things to happen.
- The development of a social life. For the time being, a range of friends rather than potential husbands. A couple of good women friends.
- Getting into physical shape.
- A hobby or avocation that I could get lost in. Something with substance.
- Some sort of volunteer work. With children, if possible.
- Some religion-related activities, not necessarily established-church related. Something that deals with the "bigger" questions of life.
- An end to all the self-recrimination over the divorce.
- Resetting my relationship with my mother (who strongly disapproved of the divorce).
- Getting over a deep-seated fear that my life is going to be bland, if not actually bleak.
- Possibly some involvement with politics.

1. Read the list from Vanessa's point of view (even though you hardly know her).
2. Since Vanessa cannot possibly do all of these at once, she has to make some choices. Put yourself in her place. Which items would you include in your agenda if you were Vanessa? In what way would the agenda items, if accomplished, help manage the overall problem situation and develop opportunities?
3. Share your package together with the reasons for each item in it with a learning partner. Discuss the differences in and the reasons for the choices. Note that there is no one right package.

My choice of possibilities for Vanessa.

_____

_____

_____

_____

_____

_____

EXERCISE 48: TURNING POSSIBILITIES INTO PERSONAL GOALS

In this exercise you are asked to review EXERCISE 43: PROBLEMS AND OPPORTUNITIES IN THE SOCIAL SETTINGS OF LIFE in which you brainstormed a range of possibilities for a better future for yourself. You are also asked to do the same for EXERCISE 44: BRAINSTORMING POSSIBILITIES FOR A BETTER FUTURE—CASES.

1. Pick two of the three social-setting you explored and two of the cases in other exercise.
2. In each, turn the possibilities that you brainstormed into one workable goal.
3. Use the criteria

a. First personal social-setting situation.

_____

b. Goal for this situation.

_____

_____

_____

a. Second personal social-setting situation.

_____

b. Goal for this situation.

_____

_____

_____

_____

a. First case.

_____

b. Goal for this case.

_____

_____

_____

_____

a. Second case.

_____

b. Goal for this case.

_____

_____

_____

_____

3. Share and critique your goals with a learning partner.

EXERCISE 49: SHAPING GOALS—MAKING SURE THAT THEY ARE WORKABLE

Accomplishing goals is a lot of work. If the goals chosen are themselves flawed, then the job becomes almost impossible. For instance, if a personal chooses a career for which he or she does not have the talent, good intentions usually can't make up for the deficit. In order to be workable, goals generally need to be:

* stated as *outcomes*,
* *specific* enough to drive action,
* *substantial* enough to make a *difference*,
* characterized by the right mixture of *risk* and *prudence*,
* *realistic*,

- *sustainable,*
- *flexible,*
- *congruent* with your *values,*
- set in a *reasonable time frame.*

These characteristics can be seen as tools that help you shape both your own and clients' goals.

1. With a learning partner return to the four goals that each of you set in the previous exercise. Use the above characteristics to review them once more. What changes might you make in each of these goals to make them conform to these characteristics?

EXERCISE 50: SHAPING—MAKING GOALS SPECIFIC

Many clients set goals that are too vague to be accomplished. "I'm going to clean up my act" is nothing more than a good intention. It is not yet a goal. Part of shaping, then, is making goals specific enough to drive action. This exercise involves moving from vagueness to concreteness.

1. First, evaluate the goal Tom and Carol in the example below have set for themselves in terms of specificity. If you don't think it is specific enough, shape it, that is make it more specific.
2. Then, in the cases that follow, get inside the client's mind and try to determine what he or she might need or want.
3. Finally, move from a vaguely stated *good intention* to a *broad aim* and then to a *specific goal.*

Example 1: Tom, 42, and his wife, Carol, 39, have been talking to a counselor about how poorly they relate to each other. They have explored their own behavior in concrete ways, have developed a variety of new perspectives on themselves as both individuals and as a couple, and now want to do something about what they have learned. In broad terms, they have agreed to stop blaming each other for their problems and take mutual responsibility for creating a better relationship. What follows is their first shot at moving from a vague to a more specific goal in the area of communication:

- **Good Intention**: "We've got to do something about the way we communicate with each other. It's just not healthy."
- **General Aim**: "We are constantly at each other over inconsequential issues and then it escalates. We'd like to do something about the way we bicker."
- **Specific Goal**: "We want to call a moratorium on what has become almost childish bickering. We want to stop second-guessing, picking on, correcting, one-upping, putting down, questioning, and being snide with each other. But we don't want to leave a vacuum. We need something positive to take its place. For instance, when we kid around, each of us should make himself or herself, not the other, the butt of any joke. When we do disagree about something important, we should avoid overly personal heated arguments and instead engage in mutual problem solving based on the helping process we are using here."

Example 2: Jeff, a trainee in a counseling psychology program, has been concerned that he does not have the kind of assertiveness that he now believes helpers need in order to be effective in consultations with clients. He is specifically concerned about the quality of his participation in the training group. First he translate his intention to be more active into a much more specific goal:

- **Good Intention.** "I need to be more assertive if I expect to be an effective helper."

128

- **Broad Aim**: "I want to take more initiative in this training group."
- **Specific Goal.** "In our group sessions, when there is relatively little structure, I want to speak up without being asked to do so. Specifically, I want to practice empathy much more."

First, critique the specific goals in both of the above examples and then work on the cases below.

1. Linda W., 68, is dying of cancer. She has been talking to a pastoral counselor about her dying. One of her principal concerns is that her husband does not talk to her about her impending death. She has a variety of feelings about dying that well up from time to time such as disbelief, fear, resentment, anger, and even peace and resignation. She also has thoughts about life and death that she has never had before and has never shared with anyone.

a. Statement of good intention.

_____

_____

b. Broad aim.

_____

_____

c. Specific goal.

_____

_____

_____

_____

2. Troy, 30, has been discussing the stress he has been experiencing during this transitional year of his life. Part of the stress relates to his job. He has been working as an accountant with a large firm for the past five years. He makes a decent salary, but he is more and more dissatisfied with the kind of work he is doing. He finds accounting predictable and boring. He doesn't feel that there's much chance for advancement in this company. Many of his associates are much more ambitious than he is.

a. Statement of good intention.

_____

_____

b. Broad aim.

_____

_____

c. Specific goal.

_____

_____

_____

_____

_____

3. Felicia, 44, finds that her lack of assertiveness is causing her problems. She is especially bothered at work. She finds that a number of people in the office feel quite free to interrupt her when she is in the middle of a project. She gets angry with herself because her tendency is to put aside what she is doing and try to meet the needs of the person who has interrupted her. As a result, she some times misses important deadlines associated with the projects on which she is working. She feels that others see her as a "soft touch."

a. Statement of good intention.

_____

_____

b. Broad aim.

_____

_____

c. Specific goal.

_____

_____

_____

_____

4. Learning partners should share the goals they have written and provide a critique for each other. When necessary, goals should be restated.

The following are supplemental exercises.
They should be filled in only if the you have not yet gotten the point or
feel that further written practice will help you master this part of the helping model.

A. Julian, 51, a man separated from his wife for seven years, has just lost a son, 19, in an automobile accident. He (Julian) was driving with his son when they were struck by a car that veered into them from the other side of the road. Julian, who had his seat belt fastened, escaped with only cuts and bruises. His son was thrown through the windshield and killed instantly. The driver of the other car is still in critical condition and may or may not live. Now, ten days after the accident, Julian is still in psychological shock and plagued with anger, guilt, and grief. He has not gone back to work and has been avoiding relatives and friends because he finds getting sympathy "painful."

a. Statement of good intention.

_____

_____

b. Broad aim.

_____

_____

c. Specific goal.

_____

_____

_____

_____

B. Joan, 32, is married and has two small children. Her husband has left her and she has no idea where he is. She has no relatives in the city and only a few acquaintances. She is talking to a counselor in a local community center about her plight. Since her husband was the breadwinner, she now has no income and no savings on which to draw.

a. Statement of good intention.

_____

_____

b. Broad aim.

_____

_____

c. Specific goal.

_____

_____

_____

_____

_____

C. Nancy, 19, unmarried, is facing the problem of an unwanted pregnancy. She has a variety of problems. Her parents are extremely upset with her. Her father won't even talk to her. She lives at home and is attending a local community college. These living arrangements are now unsatisfactory to her. Since, for value reasons, she has decided against an abortion, she does not want to live out the remaining months of pregnancy in an atmosphere of hostility and conflict. She is upset because her education is going to be interrupted and finishing college has always been high on her list of priorities. She is unsure about her finances and resents being financially dependent on her parents.

a. Statement of good intention.

_____

_____

b. Broad aim.

_____

_____

c. Specific goal.

_____

_____

_____

_____

_____

EXERCISE 51: HELPING LEARNING PARTNERS SET WORKABLE GOALS

In this exercise you are asked to act as a helper to a learning partner. This exercise is done verbally.

1. The total training group is to be divided up into smaller groups of three.
2. There are three roles: client, helper, and observer. Decide the order in which you will play each role.
3. The client summarizes some problem situation and then declares his or her **intent** to do something about the problem or some part of it.
4. The helper, using empathy, probing, and challenging, helps the client move from this statement of intent and a broad aim to a specific problem-managing or opportunity-developing goal that has the characteristics listed above.
5. When the helper feels that he or she has fulfilled this task, the session is ended and both observer and client give feedback to the helper on his or her effectiveness.
6. Repeat the process until each person has played each role.

Example: Since most students do not operate at 100% efficiency, there is usually room for improvement in the area of learning. Luisa, a junior beginning her third year of college, is dissatisfied with the way she goes about learning. She decides to use her imagination to invent a more creative approach to study. She brainstorms possibilities for a better study future, that is, goals that would constitute her new learning style. She says to herself: "In my role as student or learner, what do I want? Let me brainstorm the possibilities."

- I will not be studying for grades, but studying to learn. Paradoxically this might help my grades, but I will not be putting in extra effort just to raise a B to an A.
- I will be a better contributor in class, not in the sense that I will be trying to make a good impression on my teachers. I will do whatever I need to do to learn. This may mean placing more demands on teachers to clarify points, making more contributions, and involving myself in discussions with peers.
- I will be reading more broadly in the area of my major, psychology, not just the articles and books assigned but also in the areas of my interest. I will let my desire to know drive my learning.

She is particularly disappointed in the way she goes about writing the papers assigned in class. She believes that she wastes lots of time and ends up frustrated. And so she picks this topic for this exercise.

Declaration of intent: "I'm going to finally do something about the way I write papers for class."
Overall aim: "I'm going to put in place a systematic but flexible process for writing papers for class."

The task of the training group helper is to help Luisa develop and shape a problem-managing goal. Luisa begins by sharing her declaration of intent and overall aim. Here is some of the dialogue that follows:

JEREMY: What bothers you most about the way you have been doing your papers?

133

LUISA: Everything piles up at the end and as a result the papers are rushed and practically never represent the kind of work I'm capable of.

JEREMY: So the unsystematic approach doesn't work. Let's see what a more systematic approach might look like. What are the elements of the process you want to put in place?

LUISA: Well, I've thought about this a bit. For instance, once a paper is assigned, I will start a file on the topic and collect ideas, quotes, and data as I go along.

JEREMY: So you'll start early and have an organized filing system.

LUISA: Yes. I'll start with a single folder and just throw in ideas or quotes from books or parts of lecture notes. Ideas from anywhere or everywhere. I'll sort them out later.

JEREMY: It sounds like the sorting process could become a bottleneck.

LUISA: Hmm. You mean I'll end up with wads of stuff and wonder where to start. I know! Two things. Only one item to a sheet of paper. One idea. One quote. You know. Then after two or three weeks I'll read through and separate them into different categories and organize them in folders.

JEREMY: The sub-category approach.

LUISA: Right. Later on I can decide which topics are the most important ones. Then I can . . . .

Luisa and Jeremy go on in this vein until she has a clear idea of the process that she wants to put in place. When Luisa feels that she has a goal that will drive action, they stop and review and finetune the goal in terms of the workable-goal characteristics outlined above.

a. A summary of your problem and/or unused opportunity situation.

---

b. Your statement of intent and broad aim.

Now engage in the process outlined above.

EXERCISE 52: RELATING GOAL CHOICE TO ACTION

Once clients state what they want, they need to move into action in order to get what they want. At this point there are two ways of looking at the relationship of goals to action, one formal and one informal.

- **Formal Action.** Stage III of the helping model—brainstorming action strategies, choosing the best package, and turning them into a plan for accomplishing goals—is the formal approach to action.
- **Informal Action.** Once clients get a fairly clear idea of what they want, there is no reason why they cannot move immediately into action, that is, do *something* that will move them in the direction of

134

their goals. These are the "little" actions that can precede the formal planning process. Indeed these little actions can help clarify and finetune goals. This exercise is about these little actions.

1. State three personal goals that have emerged from your work in the training group.
2. For each goal list some "little actions" in which you might be able to engage in order to begin moving quickly in the direction of the goal.

Example. Review the case of Vanessa discussed above in Exercise 42. While she believes that it is essential for her to develop a better attitude about herself and life in general, she believes that it is not best to try to do this directly. Therefore, she chooses goals which will have a better attitude as a by-product.

- One goal centers around the development of a social life. For the time being, she wants a range of friends rather than a potential spouse. She would especially like a couple of good women friends.
- A second goal revolves around religion in some sense of that term. She would like to have religion-related activities, not necessarily established-church related, as part of her life. Something that deals with the "bigger" questions of life.
- A third goal relates to physical well being. The stress of the divorce has left her exhausted. She wants to build herself up physically and get into good physical shape.

Vanessa looks at the third goal and has this to say: "There are some things I can do immediately without coming up with a formal physical fitness program. First of all exercise. I can begin by walking—to work, to the store, and so on. Second, I can easily cut down a lot on fast food and add some fresh fruit and vegetables to my diet. Let me try these two things and see where it leads."

a. First personal goal.

_____

_____

_____

b. Actions you can take immediately.

_____

_____

_____

_____

_____

a. Second personal goal.

_____

_____

_____

b. Actions you can take immediately.

_____

_____

_____

_____

_____

a. Third personal goal.

_____

_____

_____

b. Actions you can take immediately.

_____

_____

_____

_____

_____

3. Get feedback from a learning partner on both the goals and the actions.

Section 15
STEP II-C: COMMITMENT—
What Are You Willing to Pay for What You Want?

Many of us choose goals that will help us manage problems and develop opportunities, but we do not explore them from the viewpoint of commitment. Just because goals are tied nicely to the original problem situation and initially are espoused by us does not mean that we will follow through. Even goals which seem attractive from a cost/benefit point of view can fall by the wayside. The costs of accomplishing goals might not seem too bad until it is time to pay up. Therefore, helping clients take a good look at commitment can raise the probability that clients will actually pursue and accomplish these goals. Read Chapter Fifteen before doing these exercises.

EXERCISE 53: MANAGING YOUR COMMITMENT TO YOUR GOALS

In this exercise you are asked to review the goals you have chosen to manage some problem situation with a view to examining your commitment. It is not a question of challenging your good will. All of us, at one time or another, make commitments that are not right for us.

1. Review the problem situations you have been examining in this training program and the preferred-scenario goals you have established for yourself.
2. Choose two important goals to review from the viewpoint of commitment.
3. Read the following questions and answers the ones you see relevant to each goal in order to gauge your level of commitment:

- What is your state of readiness for change in this area at this time?
- To what degree are you choosing this goal freely?
- To what degree are you choosing this goal from among a number of possibilities?
- How highly do you rate the personal appeal of this goal?
- Name any ways in which your goal does not appeal to you.
- What's pushing you to choose this goal?
- If your goals is in any way being imposed by others, what are you doing to make it your own? What incentives are there besides mere compliance?
- What difficulties do you experience in committing yourself to this goal?
- To what degree is it possible that your commitment is not a true commitment?
- What can you do to get rid of the disincentives and overcome the obstacles?
- What can you do to increase your commitment?
- In what ways can the goal be reformulated to make it more appealing?
- To what degree is the timing for pursuing this goal poor?

a. Learnings from evaluation of first goal.

_____

_____

_____

_____

_____

b. Learnings from evaluation of second goal.

_____

_____

_____

_____

_____

4.   With a learning partner, review your principal learnings from answering the above questions about these two goals. Use empathy, probes, and challenges to help one another explore levels of commitment.
5.   Finally, to the degree necessary, reformulate both goals in terms of what you have learned from the dialogue.

Restated goal #1.

_____

_____

_____

_____

_____

Restated goal #2.

_____

_____

_____

_____

_____

EXERCISE 54: REVIEWING COSTS VERSUS BENEFITS IN CHOOSING GOALS

In most choices we make there are both benefits and costs. Commitment to a preferred-scenario goal often depends on a favorable cost/benefit ratio. Do the benefits outweigh the costs? The balance-sheet methodology outlined in the Appendix of *The Skilled Helper* can be useful in determining whether goals being chosen offer "value for money," as it were. Consider the following case.

One January, Helga, a married woman with two children, one a senior in college and one a sophomore, was told that she had an advanced case of cancer. She was also told that a rather rigorous series of chemotherapy treatments might prolong her life, but they would not save her. She desperately wanted to see her daughter graduate from college in June, so she opted for the treatments. Although she found them quite difficult, she buoyed herself up by the desire to be at the graduation. Although in a wheel chair, she was there for the graduation in June. When the doctor suggested that she could now face the inevitable with equanimity, she said: "But, doctor, in only two years my son will be graduating."

This is a striking example of a woman's deciding that the costs, however high, were outweighed by the benefits. Obviously, this is not always the case. Benefits include incentives and rewards for whatever source, while costs include time and effort spent, resources used, psychological wear and tear, battling negative emotions, and the like. This exercise gives you the opportunity to explore your goals from a cost-benefit perspective. Is it worth the effort? What's the payoff?

1. Divide up into pairs, with one partner acting as client, one as helper.
2. Help your partner review one of the goals of his or her agenda from a cost/benefit perspective. The helper is to use basic empathy, probing, and challenging to help his or her partner do this. Help your partner identify benefits and costs and do some kind of trade-off analysis such as the balance-sheet technique.
3. Help you partner clearly state the incentives and payoffs that enable her or him commit to the specified goals.

Example: Carlita, 28-year-old and a third-year doctoral student in clinical psychology, is discussing her commitment to her career goals with Eban, a fellow student. She is an only child and both parents and grandparents have been pressing her to get married and have children. She wants to establish herself in her career first. She has one more year to go in the program plus a year's internship.

> **CARLITA**: If I don't get my degree and begin my career before getting married, I'll never finish. I know the kind of pressures I'd be getting from family and even friends.
>
> **EBAN**: So, from your perspective, it's almost now or never. But you're already under a lot of pressure. What about the cost-benefit equation?
>
> **CARLITA**: I'm the first woman to break the mold in my extended family. So I have to expect some costs. I don't like hearing disappointment in the voices of my parents and grandparents. But I'm setting a direction for the rest of my life.
>
> **EBAN**: You're ready to pay the price. How do you counter the pressures?
>
> **CARLITA**: First of all, I don't avoid them. I talk to them about how exciting I find my work. I talk to them about how this will benefit both me and whatever family I have in the future. Without saying it in these words, I pressure them into being proud of me, not disappointed. I'm still the same daughter and granddaughter that they have always loved. I let them know by my actions that I love them. If I gave in, that wouldn't be love. . . . In my own way I let them know how *I* need to be loved.
>
> **EBAN**: What about social life? Any pressures there?
>
> **CARLITA**: I have a couple of close men friends. But they know I'm not going to get pregnant and I'm not going to get some kind of sexual disease. Eventually, I want someone who takes me as I am.
>
> **EBAN**: So there are a few bumps there, too.
>
> **CARLITA**: None that I can't handle. I see myself as responsible and focused. If they want to see me as bullheaded, well, that's there problem. Furthermore . . . .

And the dialogue continues in this vein.

4. After ten minutes of discussion, stop and receive feedback from the client and the person being helped.

5. After the discussion, each is to get a new partner, change roles, and repeat the process.

EXERCISE 55: COMPETING AGENDAS

When setting goals, clients often fail to see them in the context of the current range of demands on their time, energy, and resources. It is no wonder , then, that some goals do not get accomplished because they are squeezed out by all the other things that the client has to do. In other words, clients have to face up to the reality of "competing agendas." Otherwise they end up trying to carry impossible burdens. Therefore, competing agendas need to be reviewed lest they dilute or drive out goals set during the helping process.

Example. A college professor, 43, is talking to a friend, who happens to be a counselor, about his values. He is vaguely dissatisfied with his priorities, but has never done much about examining his current values in any serious way. From time to time the two of them talk about values, but no conclusions are reached. He is not married. Work seems to be a primary value. He says: "Well, it's no news to you that I work a lot. There's literally no day I get up and say to myself, 'Well, today is a day off and I can just do what I want.' It sounds terrible when I put it that way. I've been going on like that for about ten years now. It seems that I should do something about it. But it's obviously my choice. I'm doing what I doing freely. No one's got a gun to my head." His major aim is to develop a social life to parallel and complement his work life. He translates this aim into a number of goals. One role centers around much more involvement in the social activities of his church, including volunteer activities. What he fails to do is to review competing agendas from the world of work. For instance, he travels internationally a great deal, something he likes, but he often returns exhausted. In the end he can't have what he wants, balance between his work life and his social life, without giving up something.

1. Look into two different goals you have set for yourself over the course of this training program.
2. Review the current commitments of your life. Outline the agendas that currently take up a great deal of your time, energy, and interest.
3. Indicate any competing agendas that might prevent you from accomplishing your goal.
4. Outline what needs to be done to manage competing agendas. This could mean dropping some current activities or reducing the scope of the goal.

a1. Competing agendas for first goal.

_____

_____

_____

_____

a2. Ways of managing competing agendas.

_____

_____

_____

_____

b1. Competing agendas for second goal.

_____

_____

_____

_____

b2. Ways of managing competing agendas.

_____

_____

_____

_____

PART FIVE

STAGE III: HELPING CLIENTS WORK FOR WHAT THEY NEED AND WANT

Stage III deals with **what** clients need to do in order to accomplish their problem-managing and opportunity-developing goals. In this stage, counselors help clients brainstorm different ways of accomplishing their goals, choose the strategies that best fit available resources, and draw up formal plans to accomplish goals.

Section 16
STEP III-A: STRATEGIES FOR ACTION—
What Do I Need to Do to Get What I Need and Want?

There is usually more than one way to accomplish a goal. However, clients often focus on a single strategy or just a few. The task of the counselor in Step III-A is to help clients discover a number of different routes to goal accomplishment. Clients tend to choose a better strategy or set of strategies if they choose from among a number of possibilities. Read Chapter Sixteen before doing the exercises in this section.

LIFE SKILLS

Sometimes people develop problems or fail to manage them very well because they do not have the kinds of **life skills** needed to handle developmental tasks and to invest themselves effectively in the social systems of life. For instance, a young married couple, Roger and Tess, find that they don't have the communication skills needed to talk to each other reasonably about the problems they have been encountering during the first couple of years of marriage. Therefore, in many cases it makes sense to help clients develop the skills they need to achieve their goals. The following is a self-assessment exercise. You are asked to review the kinds of life skills you need both to manage your own problem situations and develop unused or underused opportunities.

142

EXERCISE 56: ASSESSING CRITICAL SKILLS FOR EFFECTIVE LIVING

This exercise is a checklist designed to help you get in touch with both your resources and possible areas of deficit. Listed below are various groups of skills needed to undertake the tasks of everyday living. Rate yourself on each skill. The rating system is as follows:

5. I have a **very high** level of this skill.
4. I have a **moderately high** level of this skill.
3. From what I can judge, I am about **average** in this skill.
2. I have a **moderate deficit** in this skill.
1. I have a **significant deficit** in this skill.

You are also asked to rate how important each skill is in your eyes. Use the following scale.

5. For me this skill is **very important**.
4. For me this skill is of **moderate** importance.
3. For me this skill has **average** importance.
2. For me this skill is **rather unimportant**.
1. For me this skill is **not important at all**.

1. **Body-Related Skills**	Level	Importance
• Knowing how to eat nutritionally.	_____	_____
• Knowing how to control weight.	_____	_____
• Knowing how to keep fit through exercise.	_____	_____
• Knowing how to maintain basic body hygiene.	_____	_____
• Basic grooming skills.	_____	_____
• Knowing what to do when everyday health problems such as colds and minor accidents occur.	_____	_____
• Skills related to sexual expression.	_____	_____
• Athletic skills.	_____	_____
• Aesthetic skills such as dancing.	_____	_____

Other body-related skills:

	Level	Importance
• _____	_____	_____
• _____	_____	_____

2. **Learning-How-to-Learn Skills**	Level	Importance
• Knowing how to read well.	_____	_____
• Knowing how to write clearly.	_____	_____
• Knowing basic mathematics.	_____	_____
• Knowing how to learn and study efficiently.	_____	_____
• Knowing something about the use of computers.	_____	_____
• Using history to understand today's events.	_____	_____
• Being able to use basic statistics.	_____	_____

- Knowing how to use a library.
- Knowing how to find information I need.

   _____   _____

Other learning and learning-how-to-learn skills.

- _____

   _____   _____

- _____

   _____   _____

- _____

   _____   _____

3. **Skills Related to Values**	**Level**	**Importance**

- Knowing how to clarify my own values.
- Knowing how to identify the values of others who have a significant relationship to me.
- Knowing how to identify the values being "pushed" by the social systems to which I belong.
- Knowing how to construct and reconstruct my own set of values.

Other values-related skills:

- _____
- _____
- _____

4. **Self-Management Skills**	**Level**	**Importance**

- Knowing how to plan and set realistic goals.
- Problem-solving and opportunity-development skills.
- Decision-making skills.
- Knowing how to make incentives, punishments, and rewards work for me rather than against me.
- Knowing how to manage my emotions.
- Knowing how to delay gratification.
- Assertiveness: knowing how to get my needs met while respecting the legitimate needs of others.

Other self-management skills:

- _____
- _____
- _____

144

5. Communication Skills	Level	Importance
• The ability to listen to others actively.	____	____
• The ability to understand others.	____	____
• The ability to communicate understanding to others (empathy).	____	____
• The ability to challenge others reasonably.	____	____
• The ability to provide useful information to others.	____	____
• The ability to explore with another person what is happening in my relationship to him or her.	____	____
• The ability to speak before a group.	____	____

Other communication skills:

• _____   ____   ____

• _____   ____   ____

• _____   ____   ____

Now that you have done a brief assessment of some life skills, indicate which skills, if improved, would help you manage your concerns, problems, or soft spots better and develop opportunities more effectively. When you name a skill, indicate why such a skill is important to you and what you might do to develop it.

_____

_____

_____

_____

_____

_____

_____

What kinds of life skills do you think you need to become better at, not just to handle your own problems more effectively, but to be an effective counselor?

_____

_____

_____

_____

_____

_____

_____

EXERCISE 57: BRAINSTORMING ACTION STRATEGIES FOR YOUR OWN GOALS

Brainstorming is a technique you can use to help yourself and your clients move beyond overly constricted thinking. Recall the rules of brainstorming:

- Encourage quantity. Deal with the quality of suggestions later.
- Do not criticize any suggestion. Merely record it.
- Combine suggestions to make new ones.
- Encourage wild possibilities, "One way to keep to my diet and lose weight is to have my mouth sewn up."
- When you feel you have said all you can say, put the list aside and come back to it later to try once more.

Example: Ira, a retired lawyer in training to be a counselor, is in a high-risk category for a heart attack: some of his relatives have died relatively early in life from heart attacks, he is overweight, he exercises very little, he is under a great deal of pressure in his job, and he smokes over a pack of cigarettes a day. One of his goals is to stop smoking within a month. With the help of a nurse practitioner friend, he comes up with the following list of strategies:

- just stop cold turkey.
- shame myself into it, "How can I be a helper if I engage in self-destructive practices such as smoking?"
- cut down, one less per day until zero is reached.
- use the new nicotine-based aids, like gum, advertised on TV.
- look at movies of people dying with lung cancer.
- pray for help from God to quit.
- use those progressive filters which gradually squeeze all taste from cigarettes.
- switch to a brand that doesn't taste good.
- switch to a brand that is so heavy in tars and nicotine that even I see it as too much.
- smoke constantly until I can't stand it any more.
- let people know that I'm quitting.
- put an ad in the paper in which I commit myself to stopping.
- send a dollar for each cigarette smoked to a cause I don't believe in, for instance, the "other" political party.
- get hypnotized; through a variety of post-hypnotic suggestions have the craving for smoking lessened.
- pair smoking with painful electric shocks.

- take a pledge before my minister to stop smoking.
- join a group for professionals who want to stop smoking.
- visit the hospital and talk to people dying of lung cancer.
- if I buy cigarettes and have one or two, throw the rest away as soon as I come to my senses.
- hire someone to follow me around and make fun of me whenever I have a cigarette.
- have my hands put in casts so I can't hold a cigarette.
- don't allow myself to watch television on the days in which I have even one cigarette.
- reward myself with a week-end fishing trip once I have not smoked for two weeks.
- avoid friends who smoke.
- have a ceremony in which I ritually burn whatever cigarettes I have and commit myself to living without them.
- suck on hard candy made with one of the non-sugar sweeteners instead of smoking.
- give myself points each time I want to smoke a cigarette and don't; when I have saved up a number of points reward myself with some kind of "luxury."

Note that Ira includes a number of wild possibilities in his brainstorming session.

1. Brainstorm ways of achieving each goal. Add wilder possibilities at the end.
2. After you have finished your list, take one of the goals and the brainstorming list, sit down with one a learning partner, and see if, through interaction with him or her, you can expand your list. Be careful to follow the rules of brainstorming.
3. Switch roles. Through empathy, probing, summarizing, and challenge, help your partner expand his or her list.
4. Keep both lists of brainstormed strategies. You will use them in a later exercise.

Goal # 1.

_____

_____

_____

_____

On a separate page, like Ira, brainstorm ways of achieving this goal. Observe the brainstorming rules. When you think you have run out of possibilities, stop and return to the task later.

Goal # 2.

_____

_____

_____

_____

On a separate page, like Ira, brainstorm ways of achieving this goal. Observe the brainstorming rules. When you think you have run out of possibilities, stop and return to the task later.

EXERCISE 58:
ACTION STRATEGIES: PUTTING YOURSELF IN THE CLIENT'S SHOES

Here are a number of cases in which the client has formulated a goal and needs help in determining how to accomplish it. You are asked to put yourself in the client's shoes and brainstorm action strategies that you yourself might think of using were you that particular client.

Example: Richard, 53, has been a very active person—career-wise, physically, socially, and intellectually. In fact, he has always prided himself on the balance he has been able to maintain in his life. However, an auto accident that was not his fault has left him a paraplegic. With the help of a counselor he has begun to manage the depression that almost inevitably follows such a tragedy. In the process of re-directing his life, he has set some goals. Since his job and his recreational activities involved a great deal of physical activity, a great deal of re-direction is called for.

One of his goals is to write a book called "The Book of Hope" about ordinary people who have creatively re-set their lives after some kind of tragedy. The book has two purposes. Since it would be partly autobiographical, it would be a kind of chronicle of his own re-direction efforts. This will help him commit himself to some of the grueling rehabilitation work that is in store for him. Second, since the book would also be about others struggling with their own tragedies, these people will be models for him. Richard has never published anything, so the "how" is more difficult. For him, writing the book is more important than publishing it. Therefore, the anxiety of finding a publisher is not part of the "how."

Hobart is a graduate student in a clinical psychology training program. He puts himself in Richard's shoes and asks himself, "What would I have to do to get a book written?" Here are some of his brainstormed possibilities:

- Get a book on writing and learn the basics.
- Start writing short bits on my own experience, anything that comes to mind.
- Read books written by those who conquered some kind of tragedy.
- Talk to the authors of these books.
- Find out what the pitfalls of writing are like writer's block.
- Get a ghost writer who can translate my ideas into words.
- Write a number of very short, to-the-point pamphlets, then turn them into a book.
- Learn how to use a word-processing program both as part of my physical rehabilitation program and as a way of jotting down and playing with ideas.
- Do rough drafts of topics that interest me and let someone else put them into shape.
- Record discussions about my own experiences with the counselor, the rehabilitation professionals, and friends and then have these transcribed for editing.
- Interview people who have turned tragedies like mine around.
- Interview professionals and the relatives and friends of people involved in personal tragedies. Record their points of view.
- Through discussion with friends get a clear idea of what this book will be about.
- Find some way of making it a bit different from similar books. What could I do that would give such a book a special slant?

Now do the same kind of work for each of the following cases.

148

Case # 1. Emma, 27, has been a heroin addict. Her life has revolved around her habit. She has been told by her employer, an ad agency, that, despite her creativity, her erratic performance at work will no longer be tolerated. She opts for a new treatment—known as ultrarapid opiate detoxification—in which the drug naltrexone replaces heroin in the addict's opium receptors within six hours or so. Treatment with naltrexone is to continue for six months. Now that she is free of opiates, the challenge is to build a very different lifestyle—one that will focus her talents and energies. She is in opportunity-development mode. This includes a new social life since, currently, she has no social life to speak of. One of her goals is reconciliation with her family, including her parents, one sister, and two brothers together with their families. She has seen very little of them over the past three years apart from attending two weddings. She wants to start the work on reconciliation while she is still feeling well. She realizes that reconciliation is a two-way street and that she cannot set goals for others.

1. **Goal.** If you were Emma, what would "reconciliation with my family" look like? If accomplished what would this goal look like? What would be in place that is not now in place? How is the goal modified by the fact that reconciliation is a two-way street? Formulate a goal that has the characteristics of a viable goal outlined earlier.

_____

_____

_____

_____

_____

2. **Brainstorming Strategies.** What are some of the things you might do to achieve the kind of reconciliation with your family that you have outlined above? Do the brainstorming on a separate sheet of paper. Include some "wild" possibilities.
3. After you have developed your list, first share your goal with a learning partner. Give each other feedback on the quality of the goal. Note especially how you and your partner interpret "reconciliation."
4. Next share the lists of brainstormed possibilities. Working together, add several more possibilities to the combined lists. Keep your list for use in an exercise in Step III-B.
5. Discuss with your partner what you have learned from this exercise.

Case # 2. A priest was wrongfully accused of molesting a boy in his parish. Working with a counselor, he set three goals. His "now" goal was to maintain his equilibrium under stress. At the time other priests had been accused and convicted of pedophilia, since he knew that in the eyes of many he would be seen to be guilty until proved innocent. With the help of the counselor and some close friends, both lay and clerical, he kept his head above water. A near-term goal was to win the case in court. He also accomplished this goal. He was acquitted and all charges were dropped. After the trial the bishop wanted to send him to a different parish "to start fresh." He had been removed from his pastorate and was living at the seminary. But he wanted to return to the same parish and re-establish his relationship with his parishioners. After all, he had done nothing wrong. The bishop agreed to reinstate him in his parish.

1. **Goal.** If you were this man, what would "re-establishing my relationship with reconciliation with my parishioners" look like? What would be in place that is not now in place?? Formulate a goal that has the characteristics of a viable goal outlined earlier.

---

2. **Brainstorming Strategies.** What are some of the things you might do to re-establish your relationship with your parishioners, especially in view of the fact that some might still see you as tainted by the whole affair? Do the brainstorming on a separate sheet of paper. Include some "wild" possibilities.

Now complete steps 3-5 as above.

EXERCISE 59: HELPING OTHERS BRAINSTORM STRATEGIES FOR ACTION

As a counselor, you can help your fellow trainees stimulate their imaginations to come up with creative ways of achieving their goals. In this exercise, use probes and challenges based on questions such as the following:

- **How:** How can you get where you'd like to go? How many different ways are there to accomplish what you want to accomplish?
- **Who:** Who can help you achieve your goal? What people can serve as resources for the accomplishment of this goal?
- **What:** What resources both inside yourself and outside can help you accomplish your goals?
- **Where:** What places can help you achieve your goal?
- **When:** What times or what kind of timing can help you achieve your goal? Is one time better than another?

Example: What follows are bits and pieces of a counseling session in which Angie, the counselor trainee, is helping Meredith, the client, develop strategies to accomplish one of his goals. One of Meredith's problems is that he procrastinates a great deal. He feels that he needs to manage this problem in his own life if he is to help future clients move from inertia to action. Therefore, his "good intention" is to reduce the amount of procrastination in his life. In exploring his problem, he realizes that he puts off many of the assignments he receives in class. The result is that he is overloaded at the end of the semester, experiences a great deal of stress, does many of the tasks poorly, and receives lower grades than he is capable of. While his overall goal is to reduce the total amount of procrastination in his life, his immediate goal is to be up-to-date every week in all assignments for the counseling course. He also wants to finish the major paper for the course one full week before it is due. He chooses this course as his target because he finds it the most interesting and has many incentives for doing the work on time. He presents his list of the strategies he has brainstormed on his own to Angie. After discussing this list, their further conversations sound something like this:

ANGIE: You said that you waste a lot of time. Tell me more about that.
MEREDITH: Well, I go to the library a lot to study, with the best intentions, but I meet friends, we kid around, and time slips away. I guess the library is not the best place to study.
ANGIE: Does that suggest another strategy?
MEREDITH: Yeah, study someplace where none of my friends is around. But then I might not . . . .
ANGIE (interrupting): We'll evaluate this later. Right now let's just add it to the list.

* * * * *

ANGIE: I know it's your job to manage your own problems, but I assume that you could get help from others and still stay in charge of yourself. Could anyone help you achieve your goal?
MEREDITH: I've been thinking about that. I have one friend . . . we make a bit of fun of him because he makes sure he gets everything done on time. He's not the smartest one of our group, but he gets good grades because he knows how to study. I'd like to pair up with him in some way, maybe even anticipate deadlines the way he does.

* * * * *

ANGIE: Your strategy list sounds a bit tame. Maybe it sounds wild to you because you're trying to change what you do.
MEREDITH: I guess I could get wilder. Hmmm. I could make a contract with my counseling prof to get the written assignments in early! That would be wild for me.

Note here that Angie uses probes based on the how, who, what, when, and where probes outlined above. She also uses the "wilder possibilities" probe.

1. Divide up into groups of three: client, helper, and observer.
2. Decide in which order you will play these roles.
3. The client will briefly summarize a concern or problem and a specific goal which, if accomplished, will help him or her manage the problem or develop the unused opportunity more fully. Make sure that the client states the goal in such a way that it fulfills the criteria for a viable goal.
4. Give the client five minutes to write down as many possible ways of accomplishing the goal as he or she can think of.
5. Then help the client expand the list. Use probes and challenges based on the questions listed above.
6. Encourage the client to follow the rules of brainstorming. For instance, do not let him or her criticize the strategies as he or she brainstorms.
7. At the end of the session, stop and receive feedback from your client and the observer as to the helpfulness of your probes and challenges.
8. Switch roles and repeat the process until each has played all three roles.

Section 17
STEP III-B: BEST-FIT STRATEGIES—
What Strategies Are Best for Me?

The principle is simple. Strategies for action chosen from a large pool of strategies tend to be more effective than those chosen from a small pool. However, if brainstorming is successful, clients are sometimes left with more possibilities than they can handle. Therefore, once clients have been helped to brainstorm a range of strategies, they might also need help in choosing the most useful. These exercises are designed to help you help clients choose "best-fit" strategies, that is, strategies that best fit the resources, style, circumstances, and motivation level of clients. We begin with you.

EXERCISE 60: A PRELIMINARY SCAN OF BEST-FIT STRATEGIES FOR YOURSELF

You do not necessarily need sophisticated methodologies to come up with a package of strategies that will help you accomplish a goal. In this exercise you are asked to use your common sense to make a "first cut" on the strategies you brainstormed for yourself in Exercise 57.

1. Review the strategies you brainstormed for each of the goals considered in that exercise.
2. Star the strategies that make most sense to you. Just use common-sense judgment. Use the following example as a guideline.

Example: Let's return to the case of Ira in Exercise 57. Remember that he is the counselor trainee who wants to stop smoking. Here are the strategies he brainstormed. The ones he chooses in a preliminary common-sense scan are marked with a square (□) instead of a bullet (•).

☐ just stop cold turkey.
☐ shame myself into it, "How can I be a helper if I engage in self-destructive practices such as smoking?"
• cut down, one less per day until zero is reached.
☐ use the new nicotine-based aids, like gum, advertised on TV.
• look at movies of people dying with lung cancer.
☐ pray for help from God to quit.
• use those progressive filters which gradually squeeze all taste from cigarettes.
• switch to a brand that doesn't taste good.
• switch to a brand that is so heavy in tars and nicotine that even I see it as too much.
• smoke constantly until I can't stand it any more.
• let people know that I'm quitting.
• put an ad in the paper in which I commit myself to stopping.
• send a dollar for each cigarette smoked to a cause I don't believe in, for instance, the "other" political party.
• get hypnotized; through a variety of post-hypnotic suggestions have the craving for smoking lessened.
• pair smoking with painful electric shocks.
• take a pledge before my minister to stop smoking.
• join a group for professionals who want to stop smoking.
• visit the hospital and talk to people dying of lung cancer.
• if I buy cigarettes and have one or two, throw the rest away as soon as I come to my senses.
• hire someone to follow me around and make fun of me whenever I have a cigarette.

152

- have my hands put in casts so I can't hold a cigarette.
- don't allow myself to watch television on the days in which I have even one cigarette.
- ☐ reward myself with a week-end fishing trip once I have not smoked for two weeks.
- have a ceremony in which I ritually burn whatever cigarettes I have and commit myself to living without them.
- suck on hard candy made with one of the non-sugar sweeteners instead of smoking.
- ☐ give myself points each time I want to smoke a cigarette and don't; when I have saved up a number of points reward myself with some kind of "luxury."

4. Share with a learning partner the strategies you have starred for yourself and the reasons for choosing them.

EXERCISE 61:
BEST-FIT STRATEGIES: PUTTING YOURSELF IN THE CLIENT'S SHOES

In this exercise you are asked to put yourself in clients' shoes as they struggle to choose the strategies that will best enable them to accomplish their goals.

1. Review the case of Richard in Exercise 58.
2. Review the strategies that were brainstormed by Hobart, the clinical psychology trainee who put himself in Richard's shoes:

- Get a book on writing and learn the basics.
- Start writing short bits on my own experience, anything that comes to mind.
- Read books written by those who conquered some kind of tragedy.
- Talk to the authors of these books.
- Find out what the pitfalls of writing are like writer's block.
- Get a ghost writer who can translate my ideas into words.
- Write a number of very short, to-the-point pamphlets, then turn them into a book.
- Learn how to use a word-processing program both as part of my physical rehabilitation program and as a way of jotting down and playing with ideas.
- Do rough drafts of topics that interest me and let someone else put them into shape.
- Record discussions about my own experiences with the counselor, the rehabilitation professionals, and friends and then have these transcribed for editing.
- Interview people who have turned tragedies like mine around.
- Interview professionals and the relatives and friends of people involved in personal tragedies. Record their points of view.
- Through discussion with friends get a clear idea of what this book will be about.
- Find some way of making it a bit different from similar books. What could I do that would give such a book a special slant?

3. If you have further strategies, add them to the list now.
4. Using your common sense, star the strategies you believe belong in the best-it category.
5. Jot down the reasons for your choices.
6. Share your choices and reasons with a learning partner and give each other feedback.

EXERCISE 62: USING CRITERIA TO CHOOSE BEST-FIT STRATEGIES

Just as there are criteria for crafting preferred-scenario goals and agendas (Step II-B) so are there criteria for choosing best-fit strategies. The following questions can be asked especially when the client is having difficulty choosing from among a number of possibilities. These criteria complement rather than take the place of common sense.

- **Clarity.** Is the strategy clear?
- **Relevance.** Is it relevant to my problem situation and goal?
- **Realism.** Is it realistic? Can I do it?
- **Appeal.** Does it appeal to me?
- **Values.** Is it consistent with my values?
- **Efficacy.** Is it effective enough? Does it have bite? Will it get me there?

Example: Ira, the counselor trainee who wanted to quit smoking, considered the following possibility on his list: "Cut down gradually, that is, every other day eliminate one cigarette from the 30 I smoke daily. In two months, I would be free."

- **Clarity:** "This strategy is very clear; I can actually see the number diminishing. It gets a 6 or 7 for clarity."
- **Relevance:** "It leads inevitably to the elimination of my smoking habit, but only if I stick with it."
- **Realism:** "I could probably bring this off. It would be like a game; that would keep me at it. But maybe too much like a game. Also I know others who have done it."
- **Appeal:** "Although I like the idea of easing into it, but I'm quitting because I now am personally convinced that smoking is very dangerous for me. I should stop at once."
- **Values:** "There is something in me that says that I should be able to quit cold turkey. That has more moral appeal to me. For me there's something phony about gradually cutting down."
- **Effectiveness:** "The more I draw this action program out, the more likely am I to give it up. There are too many pitfalls spread out over a two-month period."

In summary, Ira says, "I now see that only strategies related to stopping cold turkey have bite. In fact, stopping is not the hard thing. Not taking smoking up again in the face of temptation, that's the real problem." The above criteria not only helped Ira eliminate all strategies related to a gradual reduction in smoking, but it helped him redefine his goal to "stopping and staying stopped." Sustainability is the real issue. He also noted that the brainstormed strategies he preferred referred not just to stopping but to sustainability. For instance, in the short term, the nicotine-flavored gum would reduce the craving once he has stopped.

1. Read the following case, put yourself in this woman's shoes, and, like Ira, use the criteria for determining the viability of the strategy she proposes.

Case: A young woman has been having disagreements with a male friend. Since he is not the kind of person she wants to marry, her goal is to establish a relationship with him that is less intimate, for instance, one without sexual relations. She knows that she can be friends with him but is not sure if he can be just a friend with her. She would rather not lose him as a friend. She also knows that he sees other women. She uses the above criteria to evaluate the following strategy: "I'll call a moratorium on our relationship. I'll tell him that I don't want to see him for four months. After that we will be in a better position to re-establish a different kind of relationship, if that's what both of us want."

2. Once you have done your analysis, share your findings with a learning partner. See if the two of you can learn from your differences.

**DO NOT READ THE NEXT SECTION
BEFORE DOING STEPS 1 AND 2 OF THIS EXERCISE!**

3. Now review the reasoning that the woman herself went through and her decision. Here, then, is her analysis:

*   **Clarity:** "A moratorium is quite clear; it would mean stopping all communication for four months. It would be as if one of us were in Australia for four months. But no phone calls."
*   **Relevance:** "Since my goal is moving into a different kind of relationship with him, stepping back to let old ties and behaviors die a bit is essential. A moratorium is not the same as ending a relationship. It leaves the door open. But it does indicate that cutting the relationship off completely could ultimately be the best course."
*   **Realism:** "I can stop seeing him. I think I have the assertiveness to tell him exactly what I want and stick to my decision. Obviously I don't know how realistic he will think it is. He might see it as an easy way for me to brush him off. He might get angry and tell me to forget about it."
*   **Appeal:** "The moratorium appeals to me. It will be a relief not having to manage my relationship with him for a while."
*   **Values:** "There is something unilateral about this decision and I prefer to make decisions that affect another person collaboratively. On the other hand, I do not want to string him along, keeping his hopes for a deeper relationship and even marriage alive."
*   **Effectiveness:** "When—and if, because it depends on him, too—we start seeing each other again, it will be much easier to determine whether any kind of meaningful relationship is possible. A moratorium will help determine things one way or another." Based on her analysis, she decides to propose the moratorium to her friend.

4. Discuss her reasoning with your learning partner. In what ways does it differ from your own? If you were her helper, in what ways would you challenge her reasoning and her decision?
5. Use these criteria to review the strategies you chose for yourself in Exercise 60.

EXERCISE 63:
CHOOSING BEST-FIT STRATEGIES: THE BALANCE - SHEET METHOD

The balance sheet is another tool you can use to evaluate different program possibilities or courses of action. It is especially useful when the problem situation is serious and you are having difficulty rating different courses of action. Review the balance-sheet format in the Appendix for Chapter Fifteen of *The Skilled Helper*.

Example: Rev. Alex M. has gone through several agonizing months re-evaluating his vocation to the ministry. He finally decides that he wants to leave the ministry and get a secular job. His decision, though painful in coming, leaves him with a great deal of peace. He now wonders just how to go about this. One possibility, now that he has made his decision, is to leave immediately. However, since this is a serious choice, he wants to look at it from all angles. He uses the decision balance sheet to evaluate the strategy of leaving his position at his present church immediately. We will not present his entire analysis (indeed, each bit of the balance sheet need not be used). Here are some of his key findings:

• **Benefits for me:** Now that I've made my decision to leave, it will be a relief to get away. I want to get away as quickly as possible.

    ☐ **Acceptability:** I have a right to think of my personal needs. I've spent years putting the needs of others and of the institution ahead of my own. I'm not saying that I regret this. Rather, this is now my "season," at least for a while.

    ☐ **Unacceptability:** Leaving right away seems somewhat impulsive to me, meeting my own needs to be rid of a burden.

• **Costs for me:** I don't have a job and I have practically no savings. I'll be in financial crisis.

    ☐ **Acceptability:** My frustration is so high that I'm willing to take some financial chances. Besides, I'm well educated and the job market is good.

    ☐ **Unacceptability:** I will have to forego some of the little luxuries of life for a while, but that's not really unacceptable.

• **Benefits for significant others:** The associate minister of the parish would finally be out from under the burden of these last months. I have been hard to live with. My parents will actually feel better because they know I've been pretty unhappy.

    ☐ **Acceptability:** My best bet is that the associate minister will be so relieved that he will not mind the extra work. Anyway, he's much better than I at getting people involved in the work of the congregation.

    ☐ **Unacceptability**: I can't think of any particular downside here.

• **Costs to significant social settings:** This is the hard part. Many of the things I do in this church are not part of programs that have been embedded in the structure of the church. They depend on me personally. If I leave immediately, many of these programs will falter and perhaps die because I have failed to develop leaders from among the members of the congregation. There will be no transition period. The congregation can't count on the associate minister taking over, since he and I have not worked that closely on any of the programs in question.

    ☐ **Acceptability:** The members of the congregation need to become more self-sufficient. They should work for what they get instead of counting so heavily on their ministers.

    ☐ **Unacceptability:** Since I have not worked at developing lay leaders, I feel some responsibility for doing something to see to it that the programs do not die. Some of my deeper feelings say that it isn't fair to pick up and run.

Alex goes on to use this process to help him make a decision. He finally decides to stay an extra three months and spend time with potential leaders within the congregation. He will tell them his intentions and then help them take ownership of essential programs.

1. Choose a personal goal for which you have brainstormed strategies.
2. Choose a major strategy or course of action you would like to explore much more fully. If this exercise is to be meaningful, the problem area, the goal, and the strategy or course of action in question

must have a good deal of substance to them. Using the balance-sheet methodology for a trivial issue would be a waste of time.

3.   Identify the "significant others" and the "significant social settings" that would be affected by your choice.

4.   Explore the possible course of action by using as much of the balance sheet as is necessary to help you make a sound decision.

Section 18
STEP III-C: HELPING CLIENTS MAKE PLANS—
What Kind of Plan Will Help Me Get What I Need and Want?

An plan is a step-by-step procedure for accomplishing each goal of an agenda. The strategies chosen in Step III-B often need to be translated into a step-by-step plan. The plan should include only the kind of detail needed to drive action. Overly-detailed plans usually fall by the wayside. Clients are more likely to act if they know what they are going to do first, what second, what third, and so forth. Realistic time frames for each of the steps are also essential. The plan imposes the discipline clients need to get things done. To prepare for these exercises, read Chapter Eighteen in the text.

EXERCISE 64: DEVELOPING YOUR OWN ACTION PLAN

In this exercise you are asked to establish a workable plan to accomplish one of the goals you have set for yourself. First, consider the following case.

1.   Choose a problem situation or undeveloped opportunity that you have been working on, that is, one that you have explored and for which you have developed goals and action strategies.

2.   Indicate the goal you want to accomplish.

3.   Write a simple plan that embodies the principles outlined in Chapter Eighteen.

Example. Emma, the 27-year-old heroin addict who has just undergone the ultrarapid opiate detoxification process and is not opiate-free (see Section 16), wants to establish a social life that will include family and former friends. While on heroin, she either abandoned family and friends or they abandoned her. She has explored some ways of going about this with her counselor and settles on the following plan:

*   She will contact a couple of friends and a couple of members of the family she thinks might be most receptive to her overtures. She will first write them and tell them what has happened and what she wants to do. She will follow up with a phone call.

*   Since she realizes that she might experience some rejection, she tells the members of her counseling group that she will need their support. Their support both face-to-face and by phone is an important part of her plan.

*   When she is sure of the responsiveness of both a friend and of a family member, she will arrange to meet them separately face-to-face. She will use these encounters to get the "lie of the land" and to brainstorm further possibilities.

*   She will use these initial reconciliations as a basis for further contacts. Hopefully her early contacts will also act as informal messengers to others within her family and among her former friends.

This is far as she wants to go for the moment in the planning process. She will to develop further plans only after a couple of successful contacts.

4.   Before doing your own plan, get together with a learning partner and critique the plan that Emma has developed. What are its strong points? What are its deficiencies and how might they be remedied? What might you add to the plan?

a. The goal you want to accomplish.

_____

_____

_____

_____

_____

b. On a separate sheet of paper outline the major steps of your plan.

EXERCISE 65: SHARING AND SHAPING PLANS

In this exercise you are asked to share and improve the plan you have developed and to help a learning partner do the same.

1.   Share your written plan with a learning partner. Each of you should read the other's before you get together.
2.   In a face-to-face session, explore and provide feedback to each other on the quality of the plan. How well shaped is it? To what degree is it simple without being simplistic? How concrete is it? What kind of compromise does it reach between too much and too little detail? How good is the fit between the plan and the person? Use empathy, probes, and challenge to help the other shape his or her plan.
3.   In the light of the feedback and dialogue, indicate what changes you would make in the plan.

The changes I would make.

_____

_____

_____

_____

_____

EXERCISE 66:
FORMULATING PLANS FOR THE MAJOR STEPS OF A COMPLEX GOAL

If a goal is complex, for instance, changing careers or doing something about a deteriorating marriage, the plan to achieve it will often have a number of major steps. In this case a divide-and-conquer strategy is useful. That is, the complex goal—say, the improvement of the marriage—can be divided up into a number of subgoals. For instance, a more equitable division of household chores might be such a subgoal. It is one goal in the total "package" of goals that will constitute the improved marriage.

In this exercise you are asked to spell out the action steps for two of the subgoals leading up to the accomplishment of some complex goal of your own. Consider this example.

Example: Lynette, a clinical psychology student in a counselor training program, has discovered that she comes across as quite manipulative both to her instructors and to her classmates. She believes that she has developed the style as a response to the less than enthusiastic reception she received whenever she encroached on "male" territory, whether at home, school, or in the workplace. Her current style in working with others is to make the decisions herself while letting the other party think that he or she is having a say. Then, if she does not get her way, she moves to a more overtly domineering style. However, she realizes that this style is contrary to the values she wants to permeate her relationships, including, of course, her relationships with her future clients.

Since this is the first time that she has received such feedback, one of the sub-goals she sets for herself is to get a very concrete understanding of her present style and to develop a better picture of a preferred style. This would be a major step in changing her overall manipulative style. Here are the steps in her plan to implement this style change:

- To elicit the cooperation of her instructor and the members of her training group. She wants to use the training experience as a "lab" for personal change.
- To spend three weeks interacting as she usually does. That is, for three weeks she will follow her instincts. She knows that she can't pull this off perfectly, because she is now observing herself more critically as are members of her group.
- To identify "live" instances when her domineering or manipulative style takes over. She can either catch herself or get immediate feedback from others when these instances take place. She will also keep track of instances of target behavior in her everyday life.
- To outline "on the spot," but briefly, how the interchange might have gone better. This will provide her with a practical inventory of possibilities for a change in style.
- At the end of three weeks to draw up a "portrait" of the dysfunctional style and to pull together a "portrait" of the preferred style from the possibilities developed within the group.

1. Evaluate the steps she has laid out. If she were to ask you for your help, what feedback would you give her about her plan? What changes might you suggest?
2. Now do the same for two of the major steps of a plan you have developed to achieve some complex goal related to becoming a more effective helper.

a. A key goal of mine.

_____

_____

159

b. One major step in the plan to accomplish this goal.

_____

_____

_____

c. A brief description of the steps I would take to achieve this subgoal.

_____

_____

_____

_____

_____

d. A second major step in the plan to accomplish this goal.

_____

_____

_____

e. A brief description of the steps I would take to achieve this subgoal.

_____

_____

_____

_____

EXERCISE 67: DEVELOPING THE RESOURCES TO IMPLEMENT YOUR PLANS

Plans can be venturesome, but they must also be realistic. Most plans call for resources of one kind or another. In this exercise you are asked to review some of the goals you have established for yourself in previous exercises and the plans you have been formulating to implement these goals with a view to asking yourself, "What kind of resources do I need to develop to implement these plans?" For instance, you may lack the kinds of skills needed to implement a program. If this is a case, the required skills constitute the resources you need.

1. **Indicate a goal** you would like to implement in order to manage some concern or problem situation in some way. Consider the following example. Mark is trying to manage his physical well-being better. He has headaches that disrupt his life. "Frequency of headaches reduced" is one of his goals, a major step toward getting into better physical shape. "The severity of headaches reduced" is another.
2. **Outline a plan** to achieve this goal. The plan may call for resources you may not have or may not have as fully as you would like. Consider Mark once more. Relaxing both physically and psychologically at times of stress and especially when he feels the "aura" that indicates a headache is on its way is one strategy for achieving his goal. Discovering and using the latest drugs to help him control his kind of headache is another part of his plan.
3. **Indicate the resources** you need to develop to implement the plan. For instance, Mark needs the skills associated with relaxing. Furthermore, since he allows himself to become the victim of stressful thoughts, he also needs some kind of thought-control skills. He does not possess either set of skills. Finally, he needs a doctor with whom he can discuss the kind of headaches he gets and possible drugs available for helping control them.
4. **Summarize a plan** that would enable you to develop some of these resources. In one program offered through the school, Mark learns the skills of systematic relaxation and skills related to controlling self-defeating thoughts. The Center for Student Services refers him to a doctor who specializes in headaches.

Problem situation #1

a. Your problem-managing or opportunity-developing goal.

_____

_____

_____

_____

b. Skills or other resources you need to accomplish this goal.

_____

161

_____

_____

_____

c. Summarize a plan that can help you develop or get the resources you need.

_____

_____

_____

_____

_____

Problem situation #2

a. Your problem-managing or opportunity-developing goal.

_____

_____

_____

_____

b. Skills or other resources you need to accomplish this goal.

_____

_____

_____

c. Summarize a plan that can help you develop or get the resources you need.

_____

_____

162

_____

_____

_____

_____

5. **Share** your resource-development plans with a learning partner and provide feedback to each other.

PART SIX

THE ACTION ARROW: MAKING IT ALL HAPPEN

The action arrow of the helping model in constitutes the "transition state" in which clients move from the current to the preferred scenario by implementing strategies and plans. Stages I, II, and III are all planning stages. They are all just a lot of blah, blah, blah if the client does not engage in and persist in the kind of activity that accomplishes goals. Some clients, once they have a clear idea of what to do to handle a problem situation—whether or not they have a formal plan—go ahead and do it. They need little or nothing in terms of further support and challenge from their helpers. They either find the resources they need within themselves or get support and challenge from the significant others in the social settings of their lives. At the other end of the spectrum are clients who choose goals and come up with strategies for implementing them but who are, for whatever reason, stymied when it comes to action. Most clients fall between these two extremes. There are many reasons why clients fail to act on their own behalf:

- helpers who do not have an action mentality, who help clients discuss problems, but who do not help them act.
- client inertia, that is, the failure of clients to generate the kinds of behaviors demanded by constructive change.
- client entropy, that is, the tendency of clients to slow down and give up once they have started.

Often a failure to see serious down-the-road obstacles to action contributes to both inertia and entropy. Chapter Nineteen, which deals with these and other implementation issues, should be read before doing the exercises below.

Section 19
MAKING IT ALL HAPPEN:
Helping Clients Get What They Want and Need

Since the skilled-helper model is action-oriented, it is important for you to understand the place of "getting things done" in your own life. If you are to be a catalyst for problem-managing and opportunity-developing action on the part of clients, then reviewing your own track record in this regard is important.

164

Unfortunately, helping often suffers from too much talking and not enough doing. Research shows that helpers are sometimes more interested in helping clients develop new insights than encouraging them to act on them. Inertia and procrastination plague most of us. The exercises in this section are designed to help you explore your own orientation toward action so that you may become a more effective stimulus to action for your clients.

EXERCISE 68: EXPLORING YOUR OWN ACTION ORIENTATION

Procrastination, putting things off, is part of the human condition. We all do it. In fact, we can be become quite inventive in procrastination. Some procrastination is chronic—I never seem to get around to doing papers for my courses until the last minute. But procrastination can also be incident or project specific—I know I promised to go see an ailing aunt who loves me dearly, but I don't get around to doing it. In this exercise you are asked to explore either chronic or incident-specific procrastination in your own life.

An example of chronic procrastination: Dahlia, 52, whose children, including one autistic son, are now grown, has returned to school in order to become a counselor. She has this to say about her action orientation.

> My husband is in business for himself. I take care of a lot of the routine correspondence for the business and our household. Often I let it pile up. The more it piles up the more I hate to face it. On occasion, an important business letter gets lost in the shuffle. This annoys my husband a great deal. Then with a great deal of flurry, I do it all and for a while keep current. But then I slide back into my old ways. I also notice that when I let the mail pile up I waste a lot of time reading junk mail—catalogs of things I'm not going to buy, things I don't need. All of this is odd because now that I am in school time is at a premium and I need to become more efficient.

An example of incident-specific procrastination: Randolph, a 24-year-old counselor trainee, discusses a deferred reconciliation with a cousin who attended the same college and with whom he had a falling out in his senior year:

> In high school she had been the popular one. I was a late-bloomer, I suppose, but I came to be very well-liked in college, while her social life was average. In senior year I found out that she had been telling lies about me behind my back. My success, I guess, galled her. I told her off and our relationship ended. It made family get-togethers difficult. We'd avoid each other, even though our families are close. So I found reasons to avoid family events. But from what I hear, we've both matured quite a bit the last few years. I want to re-establish a relationship with her, but I keep finding reasons to put it off.

1. Describe a form of chronic or project-specific procrastination in your own life.

_____

_____

_____

2. What are the inhibitors? What keeps from acting? What are the incentives for *not* acting?

_____

_____

_____

_____

3. What available incentives would help you move to action? What can you do to mobilize them?

_____

_____

_____

_____

4. Share your findings with a learning partner. Come up with a change program to move beyond either your habit of procrastinating or the specific instance of procrastination you described.

EXERCISE 69: EXPLORING THE SELF-STARTER IN YOURSELF

While all of us have a tendency to put things off, we also have the possibility of become self-starters. Self-starters move to problem-managing and opportunity-developing action without being influenced, asked, or forced to do so. Skilled helpers are, ideally, self-starters. They have a sense of "agency" that enable them to help clients explore self-starting possibilities in themselves.

1. Describe four ways in which you are a self-starter.

   Dahlia, whose chronic procrastination is described above, came up with the following list:

- "I never put off the things I like to do. For instance, I like the volunteer work I do at the hospital. No one has to put pressure on me to show up."
- "I make new friends easily. I don't wait for people to approach me. I take the initiative."
- "I plan family get-togethers and make them happen. Family solidarity is an important value for me. I do whatever I can to see that it happens."

- "I'm in no way a hypochondriac, but I do watch my health carefully. I get yearly check-ups, I see the dentist regularly, and if I experience symptoms that could mean something, I get myself checked out."

_____

_____

_____

_____

_____

_____

2. Describe two ways in which you would like to become a self-starter.

_____

_____

_____

_____

3. Share what you have learned about yourself with a learning partner. If you have trouble finding ways in which you are a self-starter or one to become one, explore the implications of this with the training group.

EXERCISE 70: LEARNING FROM FAILURES

As suggested in the text, inertia and entropy dog all of us in our attempts to manage problem situations and develop unexploited opportunities. There is probably no human being who has not failed to carry through on some self-change project. This exercise assumes that we can learn from our failures.

Example: Miguel kept saying that he wanted to leave his father's business and strike out on his own, especially since he and his father had heated arguments over how the business should be run. He earned an MBA in night school and talked about becoming a consultant to small family-run businesses. A medium-sized consulting firm offered him a job. He accepted on the condition that he could finish up some work in the family business. But he always found "one more" project in the family business that needed his attention. All of this came out as part of his story, even though his main concern was the fact that his woman friend of five years had given him an ultimatum: marriage or forget about the relationship.

Finally, with the help of a counselor, Miguel makes two decisions: to take the job with the consulting firm and to agree to break off the relationship with his woman friend because he is still not seriously entertaining marriage as an immediate possibility. However, in the ensuing year Miguel never gets around to taking the new job. He keeps finding tasks to do in his father's company and keeps up his running battle with his father. Obviously both of them, father and son, were getting something out of this in some twisted way. As to his relationship with his woman friend, the two of them broke it off four different times during that year until she finally left him and got involved with another man.

Some of Miguel's learnings. Since Miguel and his counselor were not getting anywhere, they decided to break off their relationship for a while. However, when his woman friend definitively broke off their relationship, Miguel was in such pain that he asked to see the counselor again. The first thing the counselor did was to ask Miguel what he had learned from all that had happened, on the assumption that these learnings could form the basis of further efforts. Here are some of his learnings:

- I hate making decisions that have serious action implications that tie me down.
- I pass myself off as an adventuresome, action-oriented person, but at root I prefer the status quo.
- I am very ambiguous about facing the developmental tasks of an adult my age. I have liked living the life of a 17-year-old at age 30.
- I enjoy the stimulation of the counseling sessions. I enjoy reviewing my life with another person and developing insights, but this process involves no real commitment to action on my part.
- Not taking charge of my life and acting on goals has led to the pain I am now experiencing. In putting off the little painful actions that would have served the process of gradual growth, I have ended up in great pain. And it could all happen again.

Notice that Miguel's learnings are about himself, his problem situations, and his way of participating in the helping process.

1. Recall some significant self-change project over the last few years that you abandoned in one way or another.
2. Picture as clearly as possible the forces at work that led to the project's ending in failure, if not with a bang, then with a whimper.
3. In reviewing your failed efforts, jot down what you learned about yourself and the process of change.
4. Share your learnings with a learning partner. Help each other: (a) discover further lessons in the review of the failed project, and (b) discover what could have been done to keep the project going.

a. **The self-change project that failed.**

_____

_____

_____

_____

b. **Principal reasons for failure.**

_____

_____

c. What you learned about yourself as an agent of change in your own life.

_____

_____

_____

_____

d. What could have been done to keep the project going?

_____

_____

_____

_____

EXERCISE 71: IDENTIFYING AND COPING WITH OBSTACLES TO ACTION

As suggested in an earlier exercise, "forewarned is forearmed" in the implementation of any plan. Identifying possible obstacles to a constructive change project is wisdom, not weakness. \

1. Picture yourself trying to implement some action strategy or plan in order to accomplish a problem-managing goal. As in the example below, jot down what you actually see happening.
2. As you tell the story, describe the pitfalls or snags you see yourself encountering along the way. Some pitfalls involve inertia, that is, not starting some step of your plan; others involve entropy, that is, allowing the plan to run into a brick wall or fall apart over time.
3. Design some strategy to handle any significant snag or pitfall you identify.

Example: Chester has a supervisor at work who, he feels, does not like him. He says that she gives him the worst jobs, asks him to put in overtime when he would rather go home, and talks to him in demeaning ways. In the problem exploration phase of counseling, he discovered that he probably reinforces her behavior by buckling under, by giving signs that he feels hurt but helpless, and by failing to challenge her in any direct way. He feels so miserable at work that he wants to do something about it. One option is to move to a different department, but to do so he must have the recommendation of his immediate supervisor. Another possibility is to quit and get a job elsewhere, but he likes the company and that would be a drastic option. A third possibility is to deal with his supervisor more directly. He sets goals related to this third option.

One major step in working out this overall problem situation is to seek out an interview with his supervisor and tell her, in a strong but nonpunitive way, his side of the story and how he feels about it.

Whatever the outcome, his version of the story would be on record. The counselor asks him to imagine himself doing all of this. What snags does he run into? Some of the things he says are:

- "I see myself about to ask her for an appointment. I see myself hesitating to do so because she might answer me in a sarcastic way. Also, others are usually around and she might embarrass me and they will want to know what's going on, why I want to see her, and all that. I tell myself that I had better wait for a better time to ask."
- "I see myself sitting in her office. Instead of being firm and straightforward, I'm tongue-tied and apologetic. I forget some of the key points I want to make. I let her brush off some of my complaints and in general let her control the interaction."

a. How can he prepare himself to handle the obstacles or snags he sees in his first statement? Then what could he do in the situation itself?

_____

_____

_____

_____

_____

b. How can he prepare himself to handle the pitfalls mentioned in his second statement? What could he do in the situation itself?

_____

_____

_____

_____

_____

Personal Situation # 1.

a. Consider some plan or part of a plan you want to implement. In your mind's eye see yourself moving through the steps of the plan. What obstacles or snags do you encounter? Jot them down.

_____

_____

_____

_____

_____

b. Indicate how you might prepare yourself to handle a significant obstacle or pitfall and what you might do in the situation itself to handle it.

_____

_____

_____

_____

_____

Personal Situation # 2.

a. Consider some plan or part of a plan you want to implement. In your mind's eye see yourself moving through the steps of the plan. What obstacles or snags do you encounter? Jot them down.

_____

_____

_____

_____

b. Indicate how you might prepare yourself to handle a significant obstacle or pitfall and what you might do in the situation itself to handle it.

_____

_____

_____

_____

_____

EXERCISE 72: FACILITATING FORCES AT THE SERVICE OF ACTION

In this exercise you are asked to identify forces "in the field," that is, out there in clients day-to-day lives, that might help them implement strategies and plans, together with forces that might hinder them. The former are called "facilitating forces" and the latter "restraining forces."

Example: Ira, as we have seen earlier, wants to stop smoking. He has also expanded his goal from merely "stopping" to "staying stopped." He has formulated a step-by-step plan for doing so. Before taking the first step, he uses force-field analysis to identify facilitating and restraining forces in his everyday life.

Some of the facilitating forces identified by Ira:

- my own pride.
- the satisfaction of knowing I'm keeping a promise I've made to myself.
- the excitement of a new program, the very "newness" of it.
- the support and encouragement of my wife and my children.
- the support of two close friends who are also quitting.
- the good feeling of having that "gunk" out of my system.
- the money saved and put aside for more reasonable pleasures.
- the ability to jog without feeling I'm going to die.
- seeing the spiritual dimensions of life.

1. Review a goal or subgoal and the plan you have formulated to accomplish it.
2. Picture yourself "in the field" actually trying to implement the steps of the plan.
3. Identify the principal forces that are helping you reach your goal or subgoal.

a. Spell out a goal or subgoal you want to accomplish and they key steps you are taking to implement it.

_____

_____

_____

b. Picture yourself in the process of implementing the plan formulated to achieve the goal. List the facilitating forces could help you to carry out the plan.

_____

_____

_____

_____

c. Which facilitating forces are most useful? What can you do to strengthen and use them? For instance, Ira focuses on three related facilitating forces: his pride, the satisfaction keeping commitments he makes, and feeling better by getting the "gunk" out of is system.

- Every morning he prayerfully recommits himself to his promise to quit.
- Every evening he congratulates himself for keeping his promises despite "flirtations with temptation."
- He periodically prides himself on becoming less dependent on chewing the gum that satisfies his craving for nicotine.
- He slowly and prudently increases his jogging time and "glories in the fresh air" in his lungs.

Now do the same for your action program.

_____

_____

_____

_____

_____

4. Finally, share your findings with a learning partner. Use empathy, probing, and challenge to help each other clarify these two sets of forces.

EXERCISE 73:
DEALING WITH RESTRAINING FORCES AT THE SERVICE OF ACTION

Now turn your attention to whatever restraining forces there might be "in the field." When these are weak, they can be brushed aside. However, when they are powerful, they must be dealt with. We return to the case of Ira.

Some of the restraining forces identified by Ira:

- the craving to smoke that I take with me everywhere.
- seeing other people smoke.
- danger times: when I get nervous, after meals, when I feel depressed and discouraged, when I sit and read the paper, when I have a cup of coffee, at night watching television.
- being offered cigarettes by friends.
- when the novelty of the program wears off (and that could be fairly soon).
- increased appetite for food and the possibility of putting on weight.
- my tendency to rationalize and offer great excuses for my failures.
- the fact that I've tried to stop smoking several times before and have never succeeded.

1. Review a goal or subgoal you identified in the previous exercise.
2. Once more, picture yourself "in the field" actually trying to implement the steps of the plan.

173

3. Identify the principal forces that are hindering you from reaching your goal or subgoal.

a. List the restraining forces that might keep you from carrying out the plan.

_____

_____

_____

_____

_____

b. Which restraining forces are most powerful? What can you do to eliminate or minimize them? For example, Klaus is an alcoholic who wants to stop drinking. He joins Alcoholics Anonymous. During a meeting he is given the names and telephone numbers of two people whom he is told he may call at any time of the day or night if he feels he needs help. He sees this as a critical facilitating force—just knowing that help is around the corner when he needs it. However, Klaus sees being able to get help anytime as a kind of dependency. He hates dependency since he sees that its roots go deep inside him. This is a major restraining force. How does he handle it?

*   First, he talks out the negative feelings he has about being dependent in this way with a counselor. That is, the first step in managing this restraining force is to name it.
*   In talking, he soon realizes that calling others is a temporary form of dependency that is instrumental in achieving an important goal, developing a pattern of sobriety. He sees calling as a safety value, not a continuation of the dependency streak within him.
*   Next he calls the telephone numbers a couple of times when he is not in trouble just to get the feel of doing so. He likes letting others know that he is in good shape.
*   He puts the numbers in his wallet, he memorizes them, and he puts them on a piece a paper and carries them in a medical bracelet that tells people who might find him drunk that he is an alcoholic trying to overcome his problem. That is, he eliminates one excuse for not using the numbers.
*   Finally, he calls the telephone numbers a couple of times when the craving for alcohol is high and his spirits are low. That is, he gets practice in using this temporary resource.

Now do the same for your change program.

_____

_____

_____

_____

_____

4. Finally, share your findings with a learning partner. Use empathy, probing, and challenge to help each other clarify these two sets of forces.

EXERCISE 74:
USING SUPPORTIVE AND CHALLENGING RELATIONSHIPS
AT THE SERVICE OF ACTION

Key people in the day-to-day lives of clients can play an important part in helping them stay on track as they move toward their goals. If part of a client's problem is that he or she is "out of community," then a parallel part of the helping process should be to help the client develop supportive human resources in his or her everyday life. In this exercise you are asked to look at strategies and plans from the viewpoint of these human resources. People can provide both support and challenge.

Example: Enid, a 40-year-old single woman, is trying to decide what she wants to do about a troubled relationship with a man. She knows that she no longer wants to tolerate the psychological abuse she has been getting from him, but she also fears the vacuum she will create by cutting the relationship off. She is, therefore, trying to develop some possibilities for a better relationship. She also realizes that ending the relationship might be the best option. Because of counseling, she has been much more assertive in the relationship. She now cuts off contact whenever he becomes abusive. That is, she is already engaging in a series of "little actions" that help her better manage her life and discover further possibilities. Over the course of two years, the counselor helps Enid develop human resources for both support and challenge.

- She moves from one-to-one counseling to group counseling with occasional one-to-one sessions. Group members provide a great deal of both support and challenge.
- She begins attending church. In the church she attends a group something like Alcoholics Anonymous.
- Through one of the church groups she meets and develops a friendship with a 50-year-old woman who has "seen a lot of life" herself. She challenges Enid whenever she begins feeling sorry for herself. She also introduced Enid to the world of art.
- Enid does some volunteer work at an AIDS center. The work challenges her and there is a great deal of camaraderie among the volunteers. It also takes her mind off herself.

Since one of Enid's problems is that she is "out of community," these human resources constitute part of the solution. All these contacts give her multiple opportunities to get back into community and both give and get support and challenge.

1. Summarize some goal you are pursuing and the action plan you have developed to get you there.
2. Identify the human resources that are already part of that plan.
3. Indicate the ways in which people provide support for you as you implement your plan.
4. Indicate ways in which people challenge you to keep to your plan or even change it when appropriate.
5. What further support and challenge would help you stick to your program?
6. Indicate ways of tapping into or developing the people resources needed to provide that support and challenge.

Summarize your goal and action plan.

_____

_____

_____

_____

_____

Identify the human resources that are already part of that plan.

_____

_____

_____

c. Indicate the ways in which people provide *support* for you as you implement your plan.

_____

_____

_____

_____

_____

d. Indicate ways in which people *challenge* you to keep to your plan or even change it when appropriate.

_____

_____

_____

_____

_____

e. What further support and challenge would help you stick to your program?

_____

_____

_____

_____

_____

f. Indicate ways of tapping into or developing the people resources needed to provide that support and challenge.

_____

_____

_____

_____

_____

TO THE OWNER OF THIS BOOK:

I hope that you have enjoyed *Exercises in Helping Skills*, Sixth Edition. I'd like to know as much about your experiences with these exercises as possible. Only through your comments and the comments of others can I learn how to make this a better book for future readers.

School: _____ Your Instructor's Name: _____

1. What I like *most* about these exercises is: _____

_____

_____

2. What I like *least* about these exercises is: _____

_____

_____

3. My specific suggestions for improving these exercises are: _____

_____

4. Some ways in which I used these exercises in class were: _____

_____

5. Some ways in which I used these exercises out of class were: _____

_____

6. Some of the exercises that were used most meaningfully in my class were: _____

_____

7. My general reaction to these exercises is: _____

8. In the space below or in a separate letter, please write any other comments about the book you'd like to make. I welcome your suggestions.

_____

_____

_____

_____

_____

9. Please write down the name of the course in which you used *Exercises in Helping Skills*.

_____

Optional:

Your name: _____ Date: _____

May Brooks/Cole quote you, either in promotion for *Exercises in Helping Skills,* Sixth Edition, or in future publishing ventures?

Yes: _____ No: _____

Sincerely,

Gerard Egan

FOLD HERE

BUSINESS REPLY MAIL

FIRST CLASS   PERMIT NO. 358   PACIFIC GROVE, CA

POSTAGE WILL BE PAID BY ADDRESSEE

ATT: *Gerard Egan* _____

Brooks/Cole Publishing Company
511 Forest Lodge Road
Pacific Grove, California 93950-9968

FOLD HERE